Montana's Pioneer N

Portrait of Morton John Elrod. August 22, 1904. (Glass negative is broken and taped together.)

Montana's
Pioneer Naturalist

Morton J. Elrod

George M. Dennison

University it y of Okl ahoma P ress : Norman

Library of Congress Cataloging-in-Publication Data
Names: Dennison, George M. (George Marshel), 1935–
Title: Montana's pioneer naturalist, Morton J. Elrod / by George M. Dennison.
Description: Norman : University of Oklahoma Press, 2016. | Includes
    bibliographical references and index.
Identifiers: LCCN 2016004183 | ISBN 978-0-8061-5436-7 (hardcover) ISBN
978-0-8061-9480-6 (paper) Subjects: LCSH: Elrod, Morton J. (Morton John),
1863–1953 | Naturalists—Montana—Biography. | College teachers—Montana—
Biography. | University of Montana—Missoula—History.
Classification: LCC QH31.E59 D46 2016 | DDC 508.092 [B] —dc23
LC record available at https://lccn.loc.gov/2016004183

The paper in this book meets the guidelines for permanence and durability of the Committee on Production Guidelines for Book Longevity of the Council on Library Resources, Inc. ∞

Rest is not quitting the busy career,
Rest is the fitting of self for its sphere.

Morton J. Elrod's philosophy, which he quoted
from Johann Wolfgang von Goethe's "Rest,"
as translated by John S. Dwight

Contents

Illustrations

Preface

Some years ago while serving as president of my alma mater, I decided to write a history of The University of Montana. After reviewing a few institutional histories, however, I nearly discarded the idea. Institutional histories tend to highlight significant dates and developments that either shaped or helped to shape the nature and structure of the emergent institutions. Frequently CEOs or other administrative figures dominate much of the narrative, which makes great sense from the perspective of explaining how and why specific key events occurred. Even so, CEOs typically exemplify variants of leadership types whose accomplishments or lack of accomplishments reflect understanding or misunderstanding of the inexorable forces at work in the historical process and institutional momentum. The nimble and celebrated leaders—through brilliant intuition or divine inspiration—correctly choose the pathway to success and remain on course at whatever price to self or others. The actions, plans, passions, and peculiarities of the many individuals who comprise the institution fade into oblivion as the institution and its builders take center stage, foreground, and background.

The most plausible explanation for this results from the available documentary evidence. Anyone who spends much time in institutional archives appreciates the impersonal nature of the documents, with some notable exceptions. During crisis periods, which occur about once every generation, personal relations and preoccupations tend to blur the institutional focus, at least briefly. Even on those occasions, the transformative leader or successful protagonist dominates but shares part of the narrative space with other participants. With the crisis resolved, institutional inertia reclaims the focus and personalities recede again from view.

As I set out to research and write the history of The University of Montana, I sought to find a way to avoid or at least to broaden the tight institutional focus. To do so, I opted to begin by capturing the perspective of a single faculty member present virtually from the founding and, if possible, trace the contributions of personal and professional development along with that of the institution. Morton J. Elrod immediately emerged as the best possible choice because of his stature at the university and the availability of his collected papers and photographs, largely unexamined by scholars. Elrod came to the university in early 1897 and took part in every major event on the campus until a paralytic stroke ended his career in 1934. More specifically, he brought to the campus his singularly unique conception of the dedicated teacher, "seldom in print" and rarely recognized, but "the power of the machine." His professional life as a faculty member illuminated the institution's trajectory from a small college on the frontier to a comprehensive undergraduate university, its modern shape and full potential still evolving, when Elrod's active career ended abruptly. He brought a restless energy and drive essential to achieving his vision for the university, for the Biological Station he founded at Flathead Lake, and for the establishment and popularization of Glacier National Park. In significant ways, Elrod's contributions and his failures added to the university's trajectory toward modernity. At the same time, he lamented the destruction that inevitably accompanied progress, even if it was at times creative destruction.

In doing the research, I contracted a huge debt to Director Donna McRae, Mark Fritch, and the staff of the K. Ross Toole Archives and Special Collections of the Maureen and Mike Mansfield Library at The University of Montana. The Elrod papers and the university archives offer rich rewards to the investigator willing to dedicate innumerable hours in search of details revealed only through extensive research. Frequently during my hours there, I thought of Professor Liston Leyendecker, an old friend who lovingly referred to his field of Western American history as "dirty fingernail history" because it required digging through primary documents. Without the assistance of dedicated staff, the task can be daunting for the researcher.

I want also to express appreciation to Charles E. Rankin, editor in chief of the University of Oklahoma Press. From his review of my first rather dreary draft, unrelieved by an almost obsessive attention to detail at the expense of the larger story, he provided the encouragement and guidance that assured, if not success, then my best approximation of it. Editorial consultants, especially Sarah C. Smith and Gary Von Euer, offered constructive counsel along the way that advised me to concentrate on readability, saved me from egregious missteps and oversights, and assisted me in making the narrative interesting as well as informative. Chris Dodge compiled the informative index. To Chuck and his expert consultants, I acknowledge my indebtedness. However, I take responsibility for any remaining errors of omission or commission.

Others gave freely of their knowledge and expertise as the book took shape. Professor William Farr, senior fellow of The O'Connor Center for the Rocky Mountain West, made early suggestions that kept me on track. Botany Professor Emeritus James R. Habeck, who has persisted in archival study of the university and its people since his retirement more than twenty years ago, generously shared the results of his research. Professor Rick Graetz followed the research and writing and afforded a willing eye and ear to test the appeal of the topic and the narrative. My colleague Mick Seidl cotaught courses with me as the book progressed and provided welcome support in innumerable ways. Director Jack Stanford of the Biological Station at Flathead Lake shared insights and provided details about Elrod's fish research and cameras. University President Royce Engstrom has continued to support the ongoing project for a history of The University of Montana.

Finally, I owe more than I can say to my wife, Jane, who has endured seemingly endless monologues about Elrod and his proclivities and idiosyncrasies. In a real way, Elrod's preoccupations became mine, and Jane found herself in the role of Emma Elrod, offering gentle reminders to enjoy life occasionally and unfailing support when challenges emerged. To Jane I dedicate this book, just as I did the first one I published nearly forty years ago. Thankfully, some things and some relationships never change.

Montana's Pioneer Naturalist

Introduction

Naturalist on the Frontier

Professor Morton J. Elrod agreed to join the small faculty of the fledgling University of Montana in 1896 to avoid, he said, becoming an itinerant professor.[1] He failed to explain a move to the far western frontier from a secure position with Illinois Wesleyan University in the bustling Midwest. His family may have fostered his westering proclivity when they moved from Pennsylvania to Iowa just after the Civil War. During the 1890s he visited several western states and succumbed to the allure of the pristine wilderness. The opportunity to work and study in the laboratory of a natural, largely unspoiled environment proved overwhelming. At the time of his final decision to make the move, the small town of Missoula formed a pastoral arcadia, untouched by the harbingers of progress and development. The relatively new state of Montana, not yet a decade old, had a population of roughly 142,900 people, not even one person per square mile; matriculated about 58 percent of its school-age children for want of schools; counted only twenty-four counties, not the fifty-six of modern times; and generated 16,800 kilowatts of power in just one hydroelectric dam. The laboratory outside Elrod's door offered a cornucopia of untold wonders for discovery.

As the seat of the university, with a population of 4,800, Missoula had become—and remained until late in the twentieth century—the state's fourth-largest town. When Elrod arrived, it had no cement or brick sidewalks, just boards and dirt. A streetcar pulled by mules went to the university, on the south side of the Clark Fork River in the old Willard School, refurbished and loaned to the university by the town. Missoula, not yet a city, to which Elrod brought his family, included 867 homes, 30 stores (including the Missoula Mercantile and the Daily Meat Packing Company), 25 saloons, and something like 20 telephones. No one had an automobile, and the roads more or less resembled trails, although the Northern Pacific Railroad had reached Missoula from Butte and Helena in 1883. The Missoula Electric Street Railway linked sections of Missoula with surrounding communities, paying fees to operate in Missoula, Ravalli, and Flathead Counties, though the streetcar did not yet go to the university's new campus at the foot of Mount Sentinel, not occupied until 1898–1899. The presence of a university, a beacon of higher learning, attested to the community's progressive ambitions.

I

In 1897, Montana incarnated what remained of the rapidly disappearing American frontier. The hardy Montanans had followed a familiar pattern, seeking admission as the forty-first state in 1889 after decades as part of an ever shrinking larger area—from the Oregon Territory in the 1850s, through Washington, Dakota, and Idaho Territories, to the Montana Territory in 1864.[2] The new state encompassed some 143,000 square miles of varied and spectacular topography, from high plains to snow-capped mountains, drained by bountiful rivers. Originating in 24,000 square miles of forest at the "crown of the North American Continent" that functioned as "an immense sponge, absorbing, holding and slowly dissipating" the water, these rivers flowed north, south, and west and poured "their mighty volumes into three oceans."[3] Initially accessible only by foot, horseback, horse and wagon, mule train, or riverboat, the area began its rapid demographic and economic growth first with the gold and silver rush of the 1860s. The growth quickened significantly after completion in 1894 of the Northern Pacific Railroad and Great Northern Railway fueled the copper boom around the flourishing metropolis of Butte.

Open-range ranching, timber harvesting, wholesale and retail business, and farming followed, all subordinate to mining and fostered by national policies to get the country's natural resources as quickly as possible into the hands of those capable of developing them.[4] The Great Northern received no federal assistance, but the Northern Pacific alone claimed some seventeen million acres of timber and rangeland in alternate sections along the route through the territory and state. Meanwhile, ranchers and homesteaders scavenged for land. Most often, they simply appropriated the land, pressing the original occupants—American Indians—onto six reservations by 1888; decades of hostile encroachment and intermittent warfare ultimately decimated the tribes.[5] At the same time, when the transcontinental railroads split the great herds, the intruders destroyed the food supply of the tribes by slaughtering the millions of buffalo native to the high plains.[6] The inevitable and tragic outcome, visible in 1883–1884, resulted in "the starvation period for the Indians." Still not satiated, the land-hungry immigrants scrambled for even more. Aided by the noble if misguided desire of reformers to civilize and protect the "savages," the settlers gained access to most of the reservations after 1887. The Dawes Allotment Act of that year, amended on two later occasions, gave the president specific congressional authorization to survey the reservations, make small allotments to individual tribal members in exchange for citizenship, and then throw open the remaining reservation land to white settlers. Ultimately, implementation of the act reduced tribal lands by some ninety million acres.[7] The terrible blizzard of 1886–1887 wreaked havoc on the settlers' unprotected cattle herds, with losses of more than 360,000 head, and ushered in a new era of enclosed ranching, woolgrowing, and homesteading.[8]

The territory and state of Montana grew in fits and starts in sync with the boom-and-bust rhythm of the unstable frontier economy, from fewer than 21,000 people in 1870, not counting thousands of American Indians, to nearly 243,300, with only 11,340 tribal members, in 1900.[9] During the 1880s, for example, the immigrant population increased by 265 percent, and the numbers continued to rise during the early twentieth century as the homestead frenzy attracted thousands of inexperienced settlers who quickly threw up fences and turned the range "grass side down," as Montana's cowboy artist Charlie Russell lamented. Towns sprouted in sheltered valleys as miners, farmers, and traders identified opportunities for gain; permanent communities began to emerge after vigilante justice eliminated known or suspected wrongdoers and a more nurturing milieu took shape.[10] However, World War I, drought and wind, and hard times forced thousands to abandon their dreams and move farther west or go back home, with a resultant population decline of more than two percent during the 1920s.[11] Between 1919 and 1925, one of every two farmers lost land, with two million acres passing out of cultivation, and farmland declined by fifty percent in average value.[12] Over those same years, 214 of the commercial banks went bankrupt, more than half of the total in the state. According to Michael Malone and Richard Roeder, these developments emphatically signaled the end of the American frontier.

Beginning in the territorial period, Montana politics trailed in the wake of national developments. The terrible economic depression of the 1890s signaled the collapse of the old village society of the nineteenth century and the emergence of the national industrial economy of the twentieth century. The depression also marked the high point of midwestern and western populism, bringing together farmers and laborers in search of federal or national restoration of order in their lives.[13] The new alliance worked for a time, not through natural affinities but because of commonly perceived oppressors. Its adherents strongly supported a federal income tax, estate taxes, and equitable corporate taxes to disrupt the concentration of wealth they saw all around them. They also fought for federal regulation of railroads and corporations; prohibition of child labor; wage and hour regulation; more direct democracy through open primaries, popular election of senators, and use of the initiative and referendum; and "free silver"—the unlimited coinage of silver—to end the economic domination of the rising industrial and financial elites.

The reform agenda initially failed in 1896 and 1900, but remnants of the farmer-labor-silver coalition merged their agenda with that of the ebullient middle-class progressive reformers. They allied to support woman suffrage, elimination of child labor, prohibition, workable and transparent municipal and state governments, and state and national social services previously provided by religious and other private or charitable groups and organizations. The new alliance persisted with a restructured agenda, although not all parts of the coalition accepted every provision of it. Nonetheless, the unstable alliance won a number of victories.[14] For a little more than a decade after the economic crisis of 1907, changing alliances persisted in Montana, typically pitting Progressive Republicans against Farm-Labor Democrats supported by the Non-Partisan League. During this period, Montana adopted woman suffrage, popular election of senators, prohibition, the initiative and referendum, tax reform, and state regulation of railroad rates and child labor. The reform surge ultimately collapsed under the pressure of war demands for patriotism and national security as well as the economic recession that struck Montana. These chaotic developments exacted a heavy toll in the state during the 1920s. The Great Depression, presaged by the market crash of 1929, brought economic disaster to the nation as a whole but merely augured more of the same for Montanans.

II

With statehood in 1889, familiar conflicts flared, sparked by the lust for the perquisites that flowed from Montana's new status: who got to host the state capitol, prisons and reform schools, and colleges and universities, and who got the positions and other political spoils. When the election in 1890 failed to resolve the issue, the advocates of Helena and Anaconda competed fiercely for the capitol in 1894, led respectively by William A. Clark, entrepreneur and future senator—briefly by purchase, then by election—and Marcus Daly of the Anaconda Copper Company. The campaign cost nearly three million dollars, roughly fifty-six dollars per vote, revealing the rapid rise of mining money influence on state politics.[15] After spending two and a half million dollars, Daly lost primarily because of the public concern about the state capitol in a town totally dominated by Anaconda Copper. Despite Daly's loss and after Henry H. Rogers and William Rockefeller acquired controlling interest of the Anaconda Copper Company in 1899, the reorganized company ruled Montana politics and maintained a copper collar on most of the daily newspapers in the state through visible as well as invisible financing.[16]

State politics during the years after 1894 turned briefly on the Clark-Daly feud until Clark and his allies won the vigorously contested election in 1900. After Daley died, Clark made peace with Anaconda Copper in exchange for the Senate seat he wanted so badly and held from 1901 to 1907. He ultimately sold his Butte mining interests to the company in 1910. In 1903, the Anaconda Copper Company closed all its operations in Butte, bringing the city and state to their knees in a display of power that forced its last serious competitor, Frederick Augustus Heinze, to concede defeat and sell his mining interests as well. Thereafter, new and younger leaders such as Joseph M. Dixon, Thomas Walsh, Burton K.

Wheeler, and Jeannette Rankin emerged and threw their leadership and support behind the eclectic reform agenda, frequently targeting the Anaconda Copper Company as the oppressor.[17]

Various communities concentrated on acquiring the statehood plums, including the colleges and university. Paris Gibson, founder of the city of Great Falls, leader of the state woolgrowers association, and future U.S. senator, pledged 360 acres of land and an endowment of $100,000 for the establishment of a single university in Great Falls.[18] Public-spirited citizens of Butte, Bozeman, Dillon, and Missoula differed, and hastened to register their claims for a school of mines, an agricultural college, a normal school, and a state university, respectively. After seeking advice from educational leaders around the country, many of whom urged a single university, the legislators debated the issue at length in 1893. But they quickly surrendered to the logrolling coalitions and distributed four campuses around the state rather than locate one consolidated university in Great Falls. Despite the fact that most states in the West chose to distribute their higher education institutions, the decision haunted state politics and the educational institutions until economic and demographic growth combined with custom to resolve the issue.[19] Even then, calls for eliminating the "limping 'colleges'" persisted.[20]

Missoula residents remained aloof from the fight over the capital, but they rallied support to bring the state university to their town. According to recollections of those involved, James M. Hamilton, the Missoula superintendent of schools, led the lobbying effort, with assistance from a group of dedicated Missoula residents and Senator Elmer D. Matts, president of the state senate.[21] Popular myth emphasizes the reliance on hard spirits, cigars, and entertainment to prevail. However, the longtime friend and colleague of Elrod, brother-in-law of Representative, Senator, and Governor Joseph M. Dixon, sometime editor of the *Missoulian,* and the university's first dean of journalism, A. L. Stone, disagreed.[22] In any event, the chartering legislation "established . . . at the City of Missoula an institution of higher learning under the name and style of 'The University of Montana,'" and authorized the State Board of Education to acquire, by purchase or gift, at least forty acres within three miles of the town for the permanent site.[23]

President James M. Hamilton and other members of the University Club persuaded E. L. Bonner and Frank Higgins to donate forty acres of grazing land across the Clark Fork River from the town center, bereft of all but scrawny shrubs and bitterroot plants. In an immediate response, the State Board of Education mandated the opening of the university in September 1895. Mary Brennan Clapp, spouse of eventual university president Charles H. Clapp, described the site as a barren plain extending to Mount Sentinel, covered with bitterroot plants and yellow bell flowers in late spring, burned brown by the sun in the fall, and "the playground of Hellgate blizzards" in winter.[24] Missoula's unpaved streets fairly boiled with dust during summer, stirred up frequently by Texas longhorn steers so formidable to encounter on the original Higgins Avenue bridge. "One old resident of the city told of having to climb over the bridge railing and hang on from outside while a herd passed," wrote Clapp.

Among their first acquaintances after arriving in Missoula, the Elrods met Mr. and Mrs. James M. Hamilton. The superintendent of Missoula public schools, Hamilton became one of Elrod's oldest and closest friends. He also served as professor of psychology and history at the university, as a member of the local executive committee for the university, and as a member of the State Board of Education. Hamilton worked during the summers with Elrod exploring Flathead Lake and Mission Mountains for several years. In 1901, he accepted a faculty appointment and then the presidency of the Montana State College of Agriculture and Mechanic Arts in Bozeman—now Montana State University—in 1904.[25]

III

Born on 27 April 1863 in Monongahela, Pennsylvania, at the height of the Civil War, John Morton Elrod, given his father's name, moved with his family to Monroe, Iowa, in 1869, foreshadowing a westering tendency that ultimately shaped his career.[26] The descendant of a long and robust line of Elrods, he reversed the order of his given names to Morton John for reasons he never explained.[27] In Iowa, he attended the public schools, taught high school during his senior year on a certificate obtained by falsifying his age, and graduated from high school in 1882. His sister reminisced years later that he actually secured a temporary permit, not a certificate, all quite legal, and that he helped support himself by working during summers on farms, and part-time during the year, by selling "stereopic [sic] views" and books. After high school, he taught public school for a time and matriculated at Simpson College, ultimately earning the BA (1887), MA (1890), and MS (1898) degrees, and then an external "Non-Resident" PhD (1905) from Illinois Wesleyan University.[28] In 1888–1889, he served as principal and teacher of the high school in Corydon, Iowa, before accepting an entry position at Illinois Wesleyan University as assistant teacher of science (1888–1889), thereafter earning promotions to Assistant Professor of Natural Science (1889–1890), and to Professor of Biology and Physics (1890–1896).[29]

At Illinois Wesleyan, he received accolades for establishing the university's science museum, and he also began his personal collection of biological specimens.[30] On 31 May 1888, he married Emma A. Hartshorn, and their union produced a daughter, Mary, born in 1889, and a son, stillborn in 1898, about a year after the move to Montana.[31] A friend extended understanding sympathy to the Elrods: "You of course learned to love the dear little one even if you never had a chance to fondle and caress it."[32] They managed the loss, as families must, consoled in the knowledge that "we have Mary, and she is worth many whole families."[33] Although of fragile health, Emma faithfully accompanied Elrod on many of his exploring excursions in their new state, joined him on location in Glacier National Park and at the Biological Station he established at Flathead Lake, and acted as secretary of the *Inter-Mountain Educator* he edited for about a decade.[34] She participated actively in the social life of their new community in Missoula and accepted his goals and aspirations, making them her own. Based on some pieces she wrote on special occasions, it appeared that Elrod influenced her rhetoric as well.[35] When apart from her, he wrote interesting and informative letters to her virtually every day. For Elrod, Emma provided a stabilizing anchor in home and family throughout their lives together.

In the 1890s before moving to Montana, Elrod made several scientific excursions to collect biological specimens in the West—primarily the Dakotas, Colorado, Idaho, and Montana—and "returned laden with rich treasures for museums," as he reported.[36] Without doubt, the irrepressible allure of the wilderness persuaded him to migrate west, although he claimed that he did so to avoid becoming an itinerant professor. He never explained what he meant by that comment, particularly in view of his apparent success at Illinois Wesleyan. Nevertheless, he exercised choice in making the move, declined the first offer he received from the

Idaho Normal School, and accepted appointment in late 1896, at age thirty-three, as the first new faculty member after the founding group at the new university in Missoula, "the prettiest town in Montana."[37] His friends thought it foolish for a man of considerable promise to waste time with this struggling institution in the remote, frontier wilderness of Montana.

Nonetheless, the Elrods departed Illinois for Montana on Ground Hog Day in February 1897. He had signed the contract with the university the prior year and resigned from Illinois Wesleyan in December 1896.[38] The board of Illinois Wesleyan University expressed its "high esteem for Mr. Elrod as a Christian man and an able instructor," predicting a bright future for him.[39] His papers provide no evidence of a noticeable religious bent, although he professed to see no conflict between religion and science, each in its own sphere.[40] Inauspiciously, the family arrived in Missoula just as the economy weakened and then tanked, presaging future but unforeseen events.

Although the Elrods had come to stay, they found the seemingly unending winter blizzards very hard to bear, especially the one in early 1899 when the wind became so strong that it blew the snow away as it fell.[41] In typical naturalist fashion, Elrod offered a scientific explanation for the winds that caused such havoc. Because of the topography of mountain ranges and deep ravines coming west from Butte and Helena and south from the high peaks north of Missoula, the cold air currents funneled down from the high pressure of the mountains into the canyons, gathering momentum with the compression into narrower, more confining spaces, and ultimately exploded through Hellgate Canyon as fierce gales. "The period of continuance may be two to five days. The wind blows constantly, never ceasing for a moment, cutting to the bone, and sometimes driving pellets of snow with such force that they sting like pebbles of stone when striking the flesh."[42] In 1899 the wind blew a foot or more of snow off the plains into the surrounding foothills.[43]

The Elrods moved into a house C. R. Prescott built in 1889, still standing today on the south corner of Gerald and Fifth, in an area that featured white picket fences around the houses to keep out the cattle. Missoulians originally called it "Bride's Row," "Mortgage Row," or "Mink Coat Row" because of the class of its occupants.[44] The surrounding communities enjoyed links with Missoula by the Missoula Electric Rail

Service, but most university students either biked or walked from where they lived. For the first few years, the university had no residence halls, and its first ones were only for female students.

At that time, Higgins Avenue extended about seven blocks south of the old bridge, where it became a country road into Pattee Canyon. As Elrod reminisced, the family lived "some ten blocks from the university campus," and "closer to the University than any other persons, faculty or students."[45] University Hall replaced the Willard School as the main facility of the university campus in 1898–99, and the county merged the Missoula High School temporarily with the university to assure a critical mass of students. Montana had only five accredited high schools in the entire state in 1895. The university charter specifically authorized a preparatory school for that reason and to assist the public schools. Because Montana as late as 1904 had only twenty-two accredited high schools, the university did not begin to phase out the prep school until 1908, even then losing the important fee revenue it generated.[46] Nevertheless, in the fall of 1908, all new students had to have earned high school diplomas, and by 1911 the prep school had fully disappeared.

In that regard, the university's entering class of fifty students in 1895 included only five who were fully qualified for college work. The remainder enrolled in the prep school, although total enrollment expanded to 135 by the close of that first school year.[47] For the first few years, some two-thirds of the students came directly to the prep school. The university faculty consisted originally of five educators: founding president and Professor of History and Literature Oscar John Craig (PhD), Professor of Natural Science S. A. Merritt (BS), Professor of Mathematics Cynthia E. Reilly (BS), Professor of Latin and Greek W. M. Aber (AB), and Professor of Languages and Mechanical Engineering Fred C. Scheuch (BME, AC). Music instructor (for fees) Mary O. Gray and librarian Mary A. Craig (BS, President Craig's daughter) rounded out the teaching staff. Aber, Merritt, and Scheuch became longtime friends and colleagues of Elrod. By 1897 when Elrod arrived, the university enrolled 176 students (57 at the collegiate level, and 44 taking music lessons alone), and the library housed 1,579 volumes. Elrod considered it significant enough to mention that, even though the university began without intercollegiate athletics, it still managed to grow without its attraction.[48]

When he accepted the challenge of establishing the department of biology and developing science education at The University of Montana, Elrod undoubtedly expected more than the minimal support he received from either the university or the state of Montana. The chance to build something from nothing while exploring a superb natural laboratory more than sufficed. In 1897, Missoula and Montana offered striking contrasts to life in other parts of the country, including the Midwest.[49] Elrod knew of the conditions because of his earlier visits, and he undoubtedly sought the position because of the opportunities presented to a naturalist in a near-pristine wilderness. Methodological naturalists such as Elrod strove to understand nature by relying on carefully controlled methods and techniques to analyze natural phenomena and events and deduce explanations that, with successful replication, earned support as natural laws.[50]

Accordingly, given the conditions within which Elrod worked and the state of scientific knowledge at the time, collecting, identifying, and classifying specimens of Montana flora and fauna constituted a great part of his research effort. He deliberately designed his doctoral dissertation, "The Butterflies of Montana," as a practical teaching aid for science instructors, and it retains its usefulness today.

The field is new. The unknown species lends zest to the search. "It isn't this and not that, it must be something new," is interesting and cheerful to hear. It shows the speaker is thinking. And the great state, with its wealth of life, beckons to the ambitious entomologist who may be the first in his locality, offering him a rich if not prolific field. To the collector will come a love for the woods and fields. They will not be places of solitude, for there he will find friends, and will commune with nature in that manner which brings the richest reward, when he is alone. He will feel the thrill of joy at first holding in his grasp a new find, for new they must be for years yet. His will be the pleasure, perhaps, of finding something new about some abundant species, for "Unknown" is yet written after many species herein mentioned. Nay, perhaps, his small collection may be the humble beginning of a larger work, limited only by physical limitations. Love for the humble little creatures of the air, love for the beautiful in nature, as revealed in their rich ornamentations, love for nature itself, with a reaction upon the individual, making him more appreciative, more happy, and more contented, will be the final reward of the young collector.[51]

Not content with description, however, Elrod always insisted on much more before conferring the honorific appellation of science. On numerous occasions, in various formulations, despite his near obsession with fieldwork, he explained that the study of science "should be accompanied by good laboratory work with suitable apparatus and material. Without this it is not real science work, and will fail of the ultimate object sought, which is *to learn by doing, to get the information first hand, and to learn to use all of the senses and correlate the information gained.*"[52] In two papers prepared for the Cosmos Club in Missoula (a town-and-gown club that still meets monthly to hear a scholarly paper or talk delivered by one of the members), he explained that the study of living things had only recently escaped the centuries-long fetters on the freedom of thought.[53] The discovery of cell theory in 1839 laid the foundation for the science of biology, and he traced the emergence of microbiology from bacteriology, beginning with the work of Louis Pasteur in the 1860s. In his opinion, the simultaneous development by Alfred Russell Wallace and Charles Darwin in 1858 of the theories of natural selection and evolution—accepted ultimately as natural laws—launched scientific biological research. He used the term "Mesology, spelled with one S," not two, as he facetiously remarked, to describe the naturalist's approach to biological science, emphasizing the influence of genetic variation in combination with adaptation to the environment for the emergence of different species.

More significantly, he included the human species within the purview of the laws governing all living things.

Man may be fattened as a pig, poisoned like an English sparrow, reduced to idiocy by a blow, . . . eats what he finds by experience he likes, as does a fish, being likewise frequently deceived, . . . responds to stimuli as does the nerveless amoeba or paramecium, . . . builds his houses by experience as derived from teaching as does the bird, the beast or the insect, maintains life and keeps warm by oxidation as do all other animals, . . . delights in his possessions and powers as does any animal, . . . and in the end mingles with the dust of the earth as does the cyclops of the pond or the giants among animals of by gone days, to be remembered only by his deeds, be they good or bad.

Even so, Elrod strove to identify the unique and natural attributes that distinguished humans from other living organisms. In notes for another Cosmos Club talk, he described "Nature Study" as the means to sharpen and strengthen the senses, the tools to understand and make use of nature and nature's laws. Trained and informed observation generated ideas or concepts that captured human curiosity and sparked the imagination to produce judgment. He defined human judgment as the active result of combining two or more ideas or concepts. In Elrod's analysis, the learning process recapitulated human evolution, enabling selection through adaptation to the environment, thus extending the benefits of human progress to all through education. While he never indicated familiarity with the Cartesian linguistic studies that led to Noam Chomsky's theories concerning development of language and mind as the divide between humans and other sentient beings, his thought pointed in that direction.[54] To support his views, Elrod cited several sources, especially Ellsworth Huntington, who strove to reveal "step by step the process by which geologic structure, topographic form, and the present and past nature of the climate have shaped man's progress; moulded his history; and thus played an incalculable part in the development of a system of thought which could scarcely have arisen under any other physical circumstances."[55] For a committed naturalist such as Elrod, the dynamic relationship between humans and their natural and intellectual environments denied determinism and aptly demonstrated human freedom within the constraints of natural law.

In that regard, John C. Beard's conclusion in 1968 that Elrod's science consisted primarily of collecting and classifying, essentially foundation work "of an exploratory nature," fails to do justice to Elrod and other methodological naturalists.[56] Elrod's love of nature informed his approach to science and deepened his obsession to read, analyze, and explain the "book of nature" in comprehensible language. Science and science education required, not merely allowed, mature scientists and students to do and see for themselves in order to grasp the inner workings of nature. His fascination with the beauty, elegance, and awesomeness of nature inspired him to poetic flights of rhetoric, including an allegorical poem entitled "The Three Seasons."[57] That same compulsion led him to require the students in his biology class to write poems on topics or creatures of their choice.

His assorted writings for the Cosmos Club and others ranged from the sublime to the mundane, with "Reason and Faith" and "Blazing the Trail"; from the mystical to the modern how-to; and from the elegiac to the practical, with "Old Places," containing the line "A real old place puts heart into a neighborhood," and "The Field Trip."[58] His notes on "Accidents to birds" observed at the Biological Station at Flathead Lake in 1900 revealed a keen appreciation for the details of living things in nature. He described a young cedar bird with no bill because of an early wound, with the result that the "tip of the tongue had become quite sharp and hard," and a young woodpecker that had lost one of its toes but had found ways to compensate.[59] In that regard, his analysis of the adaptation of land snails (*Pyramidula Strigosa*) to the high altitudes of the Montana mountains attested to his scientific acumen and knowledge.[60] In 1902, with Maurice Ricker, he reported a new species of hydra, a water worm, but the World Register of Marine Species rejected the claim because of prior discovery.[61] His writing style, while precise and well grounded in the reality he observed around him, lacked the sensuality, elegance, and passion of naturalists such as Henry David Thoreau, John Muir, and Rachel Carson. Nonetheless, his observational, analytical, and descriptive skills, combined with his acquired proficiency as a photographer, prepared him well for a career as a naturalist-educator.

In fact, Elrod established himself as an educator and an accomplished scientist among his peers. A founding member of the Montana Academy of Sciences, Arts, and Letters, he served as its first president and delivered the keynote address for the founding meeting, entitled "Montana as a Field for an Academy of Sciences, Arts, and Letters" and subsequently published in *Science*.[62] He also regularly appeared on the annual program of the Inland Empire Education Association, served as its president in 1928–1929, and accepted the onerous task of chairing the Inland Empire Science Teachers Association special committee to study science education in the public schools of the Northwest in the late 1920s. After acquiring ownership in 1914, he put the *Inter-Mountain Educator* in the service of the Montana State Teachers' Association (MSTA), teachers, and educational reformers in Montana.[63] Having become a member of the American Association of American Professors (AAUP) in 1917, he served on several of its investigating subcommittees and remained an active member until a stroke ended his career in 1934.[64]

He took organizational membership seriously and regularly attended the meetings of his chosen professional associations, such as the American Society of Naturalists, American Association for the Advancement of Science, American Microscopic Society, American Society of Zoologists, Ecological Society of America, and the Northwest Scientific Association, which he helped to found.[65] He frequently participated in the annual programs with papers, demonstrations, and comments, and he maintained a network of professional friends and associates across the country.[66] In addition, he communicated regularly with friends and peers in the Smithsonian Institution and other organizations, providing them with information about the unknown flora and fauna of Montana he helped to discover and classify. One longtime colleague and friend, Professor H. A. Pilsby at the Academy of Natural Sciences of Philadelphia, named in his honor the "*Pyramidula Elrodi*," a new species of land snail Elrod discovered in Montana and sent to him.[67] In addition, Earl Douglass, recipient of the first master's degree awarded by the university in 1899, named a fossil he discovered for Elrod, the *Procamelus elrodi*, in 1911.[68] In fact, a search of the internet for "Elrodi" produces several listings of land snails, mollusks, and the like. Elrod's contributions established and insured his status as a naturalist-scientist-educator.

His observations of the everyday world in Missoula aptly reflected his approach to science. Taking into account the available evidence, he speculated about the geological development of the Missoula region, bringing to bear a broad grounding in earth science and a keen perceptiveness. That evidence suggested a great glacial lake at least 1,000 feet deep that had covered the Missoula Valley for centuries, with beach marks on the surrounding mountains.[69] He described the Missoula Valley as rather sparse "for the study of Historical Geology, yet it contains splendid illustrations of the great dynamic forces . . . eternally at work molding and shaping the surface of the earth." He thought the excellent soil the "direct result from the mud and silt from the original lake; and the huge boulders found scattered here and there throughout the valley tell a vivid story of . . . icebergs which at one time floated on the lake's

surface . . . burdened with waste material."[70] Elrod's insightful analysis
and description preceded the extensive findings of J. T Pardee, J. Harlan Bretz, David D. Alt and their colleagues, beginning in 1910, that
ultimately established the role of Glacial Lake Missoula and its multiple
floods in forming the scablands of Idaho and eastern Washington.[71]

His account of an adventure in 1897 provides additional insight. Late
that summer, Elrod and four companions embarked on a research and
collection trip into the Flathead Indian Reservation that stretches from
just north of Missoula to about halfway up Flathead Lake and includes
portions of the majestic Mission and Swan Mountains to the east.[72] For
the trip, they loaded photographic equipment, cans, jars, chemicals,
botany presses, sleeping bags, guns, clothing, food, and other supplies
into a horse-drawn spring wagon and followed Northern Pacific's railroad track over the Marent trestle—the second highest in the world—
for thirty miles through the Mission Mountains and Valley to Flathead
Lake. They found roads better described as trails, but continued for a
considerable distance north along the east shore of the lake over a rough,
rocky stretch before turning back.

As Elrod said, the reservation boundary "is not hard to find. On one
side the ambitious white man has felled every tree of value, and has
plowed a field for grain. On the other the straight, tall, majestic pines
and tamaracks extend for miles, untouched, an unbroken wilderness.
Should the reservation . . . be thrown open for settlement, as is now
rumored, in a few years the fine timber would be gone." Confident about
the impending survey and allotment of the reservation under the terms
of the Dawes Act, Elrod subsequently took steps to secure land for a
national bison range and a biological research station.[73]

After leaving the railroad track and heading north, the party had
to struggle through giant ferns all but closing the canyons around the
St. Ignatius Mission and making the trail exceedingly difficult. Char-
acteristically, Elrod found the human specimens fascinating even if
uncollectable, and his description differed little from that of a Social
Darwinist. "Many . . . are half-breeds and quarter-breeds, who have the
wealth of the reservation, and who do work, raise the crops, build the
houses, and dig the irrigation ditches." He counted thousands of cattle
and "herd after herd of worthless cayuses," the latter sufficient "to keep
an ordinary canning factory in operation for a year, and they could be

put to no better use." The industrious tribal members had to import horses for range work. He also found it revealing that "the Mission friars, who have been getting nearly $50,000 a year from the government for running the school . . . sold 700 head of fat beef cattle, which were grazed on the reservation without cost to them, but with no 'profit' to the government" or the tribe. The Indians hunted the plentiful large game and caught the fish, but they ignored the abundant ducks, geese, and grouse.

As he interacted with the Indians, Elrod indulged his penchant for classification, grouping tribe members into three generic types. The members of the first type, referred to as "full-bloods," went about "almost naked, save for . . . [a] government blanket . . . Blear-eyed from smoke," doing virtually nothing except drawing rations and fishing. The second type consisted of the "intelligent" Indians with some education and literate in English, the "half-breeds" who physically resembled the "full-bloods" but shared the "inventive genius" and "aggressiveness" of whites, who expertly managed their farms and ranches and took their products to market for sale. Finally, the "quarter-breeds" did not look like Indians at all and exemplified the attributes less developed among the "half-breeds." Elrod and his companions spent a pleasant evening for dinner with one family of the latter type, entertained by the host's daughter on the organ and son on the violin. Elrod concluded optimistically that the large but unknown number of thrifty Indians similar to their host demonstrated conclusively that "The tendency is upward. The white man is slowly but surely amalgamating the Indian," with government policy supporting that objective. Some people he consulted, including Indian agent Major J. W. Powell, argued that full-bloods no longer existed. Elrod predicted that soon "the lazy Indians will all be dead, and the thrifty Indians will not need a reservation," thereby justifying the imminent allotments and his own intention to claim land.

Elrod's concern for all living things moderated the obvious racial and cultural bias of his time—demonstrated in the preceding paragraph—and inspired him to learn about Indian life and culture. More importantly, it encouraged him to expand and preserve the collection of cultural artifacts, myths, and various stories to enhance public understanding of American Indians more broadly.[74] An encounter in the Mission Mountains with three Indians during a severe rainstorm stimulated some revealing comments.[75] With a gesture typical of his respectful attitude, Elrod provided food, matches, and some warm clothing to the Indians. As he explained, "It has always been my custom to give freely of eatables or other little things to Indians when there was occasion to render service. . . . And I always found them responsive and appreciative. I was in their domain. I was not supposed to be there. True, I had a paper . . . from the agent, but . . . I was in their country." He traversed the reservation frequently and always complied when asked by the Indian police to show his permit.

On occasion, he compared the fate of the Indian to that of the buffalo, expressing deep and sincere concern for both, and the corrosive effect of modern civilization on the Indian's distinctive way of life. In addition, he proposed strengthening the university museum's existing but inadequate collection in order to illuminate the former life of Montana Indians.[76] He thought a larger, more comprehensive effort necessary, even if that required more space and support for the museum. He also devoted a Biological Station bulletin to the "Pictured Rocks, Indian Writings on the Rock Cliffs of Flathead Lake, Montana."[77] With J. M. Hamilton, Maurice Ricker, and J. P. Rowe, he studied the site of the "Writings" at Angel Point and prepared technical descriptions of the area and the pictographs, but he respectfully declined to attempt an interpretation. The press coverage and the bulletin informed the scholarly community and the public about this lasting vestige of an earlier civilization. Years later, Robert G. Raymer, an anthropologist at Carroll College, requested a copy of the bulletin on the pictographs, by then out of print.[78]

Elrod's passion for nature and all living things and his love of the Montana wilderness manifested a powerful impulse toward conservation and preservation. Over the years, he formed associations with leading figures in the emerging American conservation movement, including Theodore Roosevelt, William Temple Hornaday, and George Bird Grinnell, among others, and he joined them in a number of projects.[79] Grinnell, the recognized "father of American conservation," called for the protection of the headwaters of the great rivers originating in Montana for the benefit of the entire North American continent, a call that Elrod strenuously echoed.

Elrod's acute awareness of the plight of Indians and buffalo and other animal and plant species reinforced this naturalist tendency. As for the

2,200 Indians inhabiting land on the Flathead Indian Reservation after allotment, Elrod's sentiments strongly colored his wife Emma's conclusion that "the advent of civilization" doomed them and their picturesque way of life to oblivion.[80] He himself wrote plaintively of the rapid disappearance of Montana's native birds and animals as well and assiduously pursued his collecting activity in a race with time.[81] He dreaded the onslaught of animal and bird extinctions symbolized by the slaughter of the buffalo and the uncontrolled hunting of the grizzly bear.[82] The antelope, he predicted, "will likely be saved only by heroic measures" from the polluted rivers, fences enclosing formerly open areas and interrupting animal pathways, and savage destruction of the animal and its natural predators. On several occasions he warned of the fragility of the rapidly vanishing wilderness: "The wonderful primeval forests have been almost entirely removed, and grain fields, gardens and orchards substituted. Holt, the post office [near Bigfork], has long since passed. The beautiful woods of Swan Lake have been logged, and a town now stands where a few years since our camp was undisturbed save by wild birds and beasts."[83]

In words and passages calling to mind the nineteenth-century painter Thomas Cole's "Course of Empire" sequence depicting the decay that inevitably followed in the wake of development, Elrod gave voice to a deep yearning for a disappearing arcadia that influenced him throughout his career.[84] As he wrote in support of but prior to the designation of Glacier National Park, "The works of nature are not marred by the hand of man, save for an occasional trail." He concluded, "A Park like this will not only preserve the natural scenery from ruination by civilized (or uncivilized) methods, but its creation is the only logical way of saving to humanity the many forms of life fast becoming extinct." A similarly strong moral imperative induced him to accept without hesitation an invitation to represent the western states on the Ecological Society of America's Committee on the Preservation of Natural Conditions for Ecological Study.[85] With his peers and colleagues in the nascent conservation-preservation movement, Elrod exhibited deep ambivalence about the desired end, whether the conservation of resources for future use or the preservation of the pristine natural state.[86] Elements of that ambivalence continue today in the environmental movement.

IV

Having accepted a faculty position in a public institution located in a relatively poor frontier state, Elrod had to struggle to maintain a decent standard of living for his small family. In that regard, his monthly salary averaged less than $200 until some point in the 1920s, roughly $27,000 a year in today's dollars.[87] After all, he had moved to Montana to avoid becoming an itinerant professor, not to live in poverty.[88] Over his active career, he seized every opportunity for reasonable gain that presented itself. Thus he received "several hundred dollars," including expenses, for identifying a site in Montana for the National Bison Range established by the American Bison Society in 1907. Documenting the experience, he also published two articles that received national attention and elevated his scholarly standing.[89] Up until 1910, he received only a few dollars for expenses at the Biological Station, and thereafter only about $200 as salary. In 1906, he contracted as an expert witness for the plaintiff in litigation alleging damage from the smoke and toxic pollution spewing from the massive Anaconda Copper Company Washoe Smelter in Anaconda. During those same years, he marketed postcards made from his photographs (6.5 in. by 8.5 in.) either by direct retail sale or by mail at fifty cents each. His customers included people unable to see the Indians in person or who simply wanted mementos, including shots of the annual powwow and other gatherings and dances, and of Duncan McDonald, the Salish elder.[90] In 1910, he wrote to Emma reporting sales of nearly a thousand cards in Ravalli and had yet to test the market in Polson.[91]

After 1900, Elrod also launched lecture tours around the state and region.[92] During the decade of the twenties, he spent the summers as a naturalist in Glacier National Park at a prorated monthly salary starting at $100 plus expenses in the park. While developing the park's Nature Guide Service, he wrote, published privately, and marketed an authorized tourist's guide to the park that generated revenue well into the 1930s.[93] In 1908, he published privately a booklet of photographs featuring twenty plates and a multiplate panorama of the Mission Mountains to depict some of the "most sublime scenery in America."[94] Seven years earlier, he published a piece with a few photographs to describe the rush of sensual pleasure upon coming over the rise from Ravalli with the full splendor of the Mission Mountains spread before him, a truly memorable experience.[95]

In addition, he conducted a brisk business during the 1920s selling seeds of flowers and other plants in the park on site or by mail order, having discovered even earlier a market for tree seeds.[96] In the years after 1910, he purchased stock in a cattle company in eastern Montana and bought at least five lots on Flathead Lake for $125 each ($10 an acre) for resale.[97] Beginning in 1914 and for a decade, he owned, published, and edited the *Inter-Mountain Educator* for the MSTA at an annual profit most years.[98] Finally, Elrod published for his friend, Arthur L. Stone, editor of the *Missoulian* until he became the founding dean of the university's School of Journalism, the volume of stories Stone wrote initially for the *Missoulian*.[99] No record of the division of the proceeds he shared with Stone exists, but the book generated a significant amount of money at the time and later in the twentieth and twenty-first centuries.[100]

V

For nearly four decades, until a paralytic stroke ended his career as a naturalist-scientist-educator, Elrod labored to take full advantage of the opportunities available to him and to conserve and preserve as much of Montana's natural splendor as possible. Without question, he enjoyed the frontier years immensely, and reminisced wistfully about them as the years passed. His work also earned the respect and admiration of his peers and the appreciation of his students because of the value they derived from studying under his direction. One former student, Harold C. Urey, wrote upon receiving the Nobel Prize in Chemistry that he attributed his success to Elrod's influence and the ambience at The University of Montana. "Nobel Prize winners come from small schools," he hypothesized, because they are nurtured by the attention they receive. "It pushes up their vanity, their self-regard and induces them to do an enormous amount of work which otherwise they might not do."[101] Urey aptly described the relationship between this distinguished naturalist-educator and his disciples.

CHAPTER ONE

A University, a Biological Station, and a Bison Range

Elrod accepted the challenge to build a biology department at The University of Montana even though he had only a decade of experience in a private, parochial, and well-funded university. After earning his Bachelor of Arts degree in 1887, he secured his first university faculty position and moved quickly through the ranks, from assistant instructor of biology in 1888 with a Bachelor of Arts degree, reflecting a broad focus in the liberal arts and sciences, to assistant professor of natural science in 1889. He received his Master of Arts degree in 1890 and became professor of biology and physics in 1891.[1] In 1896, he accepted appointment as professor and chairman of the department of biology at Montana, and received his Master of Science degree in 1898, with little if any additional didactic work. To earn the PhD, he fulfilled the reading, laboratory, and examination requirements of the first external degree program offered in the United States, and, with no further formal course work, prepared and submitted a dissertation accepted by Illinois Wesleyan University to receive the degree in 1905.[2]

Elrod's access to the doctorate followed a route unusual in both the United States and Europe. By way of comparison, Albert Einstein submitted in 1905 one of four papers he had recently written as his dissertation to receive his doctorate from the University of Zurich in Switzerland, after an earlier submission was rejected in 1901.[3] Einstein's dissertation on the size of molecules and his other work changed the world of science. On the other hand, Elrod pragmatically identified and classified the butterflies of Montana and included photographs and illustrations based on fieldwork in Montana between 1897 and 1898 and on analyses of existing collections.[4] While not revolutionary in its impact, his work has withstood the test of time. With dedication and effort, Elrod sought and obtained the degrees he found useful in the work and positions he undertook, a familiar pattern during his lifetime and even in the twenty-first century.

In that regard, most of Elrod's colleagues at the new university on the frontier never bothered to obtain advanced degrees, since they primarily taught courses for the university prep school or for undergraduate students, as Elrod did. Although The University of Montana awarded its first graduate degree—the Master of Science—to Earl Douglass in 1899, graduate education claimed little attention until well after Elrod retired.[5] In fact, by 1942 the state university had conferred only 256 master's degrees and no doctorates, with education degrees accounting for the largest number (75). Elrod's close friend and colleague, professor and chairman of the chemistry department William Draper Harkins, followed nearly the same pattern. However, Harkins skipped the master's degree and took leave without pay from Montana for at least two terms in residence prior to completing the doctorate in chemistry at Stanford University, awarded eight years after he came to Montana as professor and chairman with a BS degree.[6] Harkins produced a dissertation that reported the chemical analyses he and his mentor conducted under contract to demonstrate the effects of smelter smoke on livestock in the Deer Lodge Valley.[7]

As a faculty member, Elrod engaged in a diverse array of activities beyond his heavy instructional and other university assignments, activities that varied with circumstances and opportunities and broadened his experience and knowledge of the world around him. At different times, his activities reflected his dedicated interest in science and nature, his entrepreneurial efforts to increase his income through congenial work, or his passionate concern for the conditions and challenges in higher education. As a naturalist-educator, he found ways to remain engaged and active while always alert to the main objective. He typically exhibited a strong sense of community responsibility as well as self-confidence, which drove him to engage in projects he judged of immediate social and personal benefit, although he rarely ventured into political activism. Even while pursuing his own interests, he took seriously the obligations of a committed citizen and dedicated himself to fulfilling them. He often described himself as utterly honest and reliable, possessed of no little talent, willing to work hard for the attainment of valued goals, and committed to the best interests of the university, Missoula, and Montana, and his friends agreed with that assessment.[8] Yet others were offended by the way he argued his opinions about the need for high standards, his tendency toward self-righteousness, and his frequent lapses of discretion when engaged in discussion. In fact, his obdurate stands on principle nearly cost him his university position.

I

Elrod's work in the biology department began in February 1897 with gratifying results, according to President Oscar J. Craig, with whom he had a good and respectful relationship.[9] As did many after him, James M. Hamilton credited Craig with building the university from scratch, devising ways to initiate instruction in 1895 without adequate state support, and financing the construction of the first facilities on the barren tract of land that became the site for the state university.[10] The university's charter authorized a preparatory department because of the paucity of accredited high schools in the state, and Elrod strongly supported Craig in working with the public high schools. To that end, he provided staff support for the State Board of Education special committee (composed of J. M. Hamilton, J. W. Kleck, Oscar J. Craig, and D. W. Sanders) charged to design the curricula required for high school accreditation.[11] Nonetheless, differences between the two men about policy matters soon surfaced.

Craig's initial plan for high school accreditation included three curricula—classical, science, and English—and schools meeting the requirements of one or all three earned accreditation by the state board, thereby assuring their graduates admission to the university without examinations.[12] Elrod objected, convinced that the abandonment of admission examinations lowered academic standards because it allowed high schools to decide who got admitted by certifying their graduates. He thought the prep school already opened the campus to unprepared and unmotivated students, though he understood the reason for it.[13] Making matters even worse, the faculty of the university voted to accept any course offered by an accredited high school. Elrod fumed that this irresponsible action, approved by Craig, exacerbated the originally flawed approach.[14]

This disagreement in principle soon transformed Elrod into an outspoken advocate for the elimination of the prep school and the maintenance of university admission examinations. With rising frustration, he waited impatiently for the closure of the prep school. As a solution to the problem of unprepared students, Elrod advocated standardized admission examinations by an external agency, such as the Scholastic Aptitude Test (SAT) or American College Test (ACT) then in development. He ultimately endorsed physical and mental examinations to establish individual ability and to require a program of study tailored to assure academic success: "No one has a right to ask or expect more, and the discovery of ability must be possible, no matter what may be the objection or protest."[15]

Undeterred, President Craig subsequently added a fourth approved high school curriculum for prospective business students, offering one more route to admission without examination.[16] In addition, he obtained board reconfirmation of the accreditation and admission processes, and recommended a high school board to consist of the four institutional presidents and the state superintendent of schools. The new board would monitor the high schools through an inspector it appointed.[17] But the state board rejected this recommendation, so Craig, as university president, continued to serve as inspector of high schools until the state board later assigned that responsibility to the state superintendent of schools.[18]

Nonetheless, in 1907 Elrod agreed when Craig stridently condemned the state legislature's refusal to authorize State Board of Education certification for those university graduates seeking to become public school teachers. Both men regarded this refusal as a serious error damaging to public education in the state.[19] Apparently Elrod thought universities more principled and reliable in the certification of their graduates than high schools. Despite this temporary setback, Craig's commitment to the public schools continued undiminished.

Never willing to compromise on principle, Elrod's disdain for unmotivated and underprepared students soon generated public differences with the president. His rather cryptic comment in 1905 to Emma about receiving his doctorate hinted at a deteriorating relationship with Craig:

"Did I tell you I got my degree diploma before leaving home? I did not tell the prexy, don't intend to until next catalogue time. He would not appreciate it."[20]

As a further source of irritation to the president, Elrod persisted in placing orders for equipment and supplies directly with vendors, ignoring the president as the university purchasing agent.[21] Disgusted by Elrod's insubordination, Craig threatened vendors with nonpayment for violation of university rules.

An annoying incident in 1898 sparked by Elrod's first report on the Biological Station, the comment about the dissertation, and the trouble with vendors signaled that Elrod, and perhaps his colleagues as well, soon developed a strained relationship with President Craig. Although he authorized the constitution and bylaws drafted by a committee—with Elrod as a member—that permitted limited faculty involvement in governance, Craig presided in the manner of the old-time college president, according little more than gratuitous condescension to faculty opinion.[22] Professor William Aber, a member of the founding faculty, commented that the president's sensitivity about his dignity and executive prerogatives made open discussion of serious issues nearly impossible.[23] However, this deteriorating relationship remained beneath the surface until it erupted in 1908, nearly causing disaster for Elrod.

II

Founding president Oscar John Craig built a campus from nothing, with virtually no state-appropriated funds and in the face of persistent public criticism for trying to support four separate campuses rather than just one. In response, Craig argued passionately that "the course of Higher Education demands not so much consolidation of schools and colleges as their proper adjustment. Let each be employed in its own work . . . within the limits set by the statute."[24] He saw no need to consolidate the separate campuses if his colleagues heeded his counsel. To that end, he pledged to manage the university as a business proposition to achieve desired results. On various occasions, he pleaded for a stable fiscal policy, one not subject to abrupt change, to assure permanent support for the university. Under such an arrangement, he committed the university to prepare Montanans for positions of honor and trust. Unquestionably, an annual budget of $20,600 failed that criterion miserably. His lack of success in securing the resources necessary to support Elrod's persistent requests inevitably tested their relationship and darkened the president's mood. Oblivious to the signals that others perceived, Elrod continued to press for his priorities.

Compounding his negativism, Elrod soon found that managing prep school students and regularly admitted undergraduates in the same room exacerbated the usual challenges of teaching.[25] His instructional assignments at Illinois Wesleyan had provided no experiential foundation for him. Space, equipment, and personnel constraints further frustrated his efforts. He finally persuaded the president to acquire some new scientific equipment that helped immensely, as did the student field trips to Mount Sentinel and to Rattlesnake, Pattee, and Hellgate Canyons. Department research and investigations by the students also improved markedly in quality.[26] Elrod and other scholars launched studies during these early years at the Biological Station at Flathead Lake that Elrod founded for research purposes in 1898. At the same time, he began to urge a school of forestry within the biology department, identifying the Biological Station and the O'Brien mill near Somers as ideal sites to train students to manage timber production, manufacture lumber, and pickle railroad ties.[27] He thought a forestry school offered a way to strengthen and diversify the department's work while at the same time responding to the state's need to manage its renewable resources. He anticipated rapid development of the timber industry in Montana, sooner rather than later, because he doubted the sustainability of forests in the East under the current harvesting regimen, an accurate prediction.

The university enjoyed steady but moderate growth in enrollments and took incremental if small steps to provide needed facilities. President Craig reported campus enrollments of 347 students in 1902, up to 393 by 1907.[28] In 1902, the state board issued bonds to construct a residence hall on campus for females and a gymnasium for physical education and recreation.[29] Craig and most of the faculty considered residence halls for men superfluous, an anachronism left by former military schools. Some of the funds also went to repair fire damage to Science Hall, constructed a year earlier.

As the final construction projects during his tenure, President Craig secured authorization for a university library and for renovation of the heating plant in 1907.[30] Thereafter, declining state revenues and the advent of World War I delayed most new construction until after 1920, when the voters approved a new funding approach for public higher education. However, Representative Joseph M. Dixon and Senator Paris Gibson secured a federal grant of 480 acres on Mount Sentinel for an astronomy observatory (unfortunately never completed), and President Craig persuaded the Northern Pacific Railroad to release its claim to an adjoining forty acres east of campus.[31] In addition, the president planned and executed the construction of athletic fields in the northeast sector of the campus, funded with fee revenue, a project that enhanced intercollegiate athletics but introduced new conflicts over the role of athletics. While he supported intercollegiate athletics, Elrod never understood allowing auxiliary projects to preempt academic facilities.[32]

Even though by 1902 a majority of the entering students each year qualified for collegiate admission, Craig steadfastly continued the prep school.[33] He valued it not only because of the need for enrollments and fee revenue but also because it assured access for more Montanans while at the same time providing a model for emulation by the public schools. Quoting the president of Cornell University, he agreed that "no community can long maintain a system of common schools worthy of the name without a college or university above." Clinging fiercely to that vision, he alienated an increasing number of the faculty who shared Elrod's views. In Craig's opinion, however, "The influence of the University in strengthening and unifying the public school system of Montana has been very marked and is becoming more and more apparent."[34]

Claiming sufficient progress in 1905, Craig finally recommended the closure of the prep school to occur over three years after 1908, with admissions and instruction reduced by one third annually. At the same time, he expressed great pride in his accomplishments and asserted that "in no case is the influence of the University upon the schools arrogant or dictatorial. It partakes rather of the spirit of the new ethics which emphasizes the obligation of the strong to help the weak, of the higher to the lower." Craig's commitment to *noblesse oblige* struck Elrod as little more than an abject abandonment of academic standards. Critically, he openly expressed his views with increasing intensity.

III

Yet Elrod also sought to work with the president, despite their evolving differences. Years later, with more experience and time for reflection, he praised Craig's unselfish dedication to the university. Even the students agreed, as in their meetings they recited and institutionalized Craig's famous epigram "the University of Montana, it must prosper."[35] Elrod also named Craig Mountain in the Mission Mountains to honor the president's unceasing support for the university and the Biological Station.[36] In 1921, he recalled his first outreach effort when he participated in the founding meeting of the Montana Horticulture Society at Craig's request in 1897.[37] Although claiming modest expertise in botany, he remained active in the Horticulture Society, serving several years as an officer and attending meetings into the 1920s. His interest reflected a commitment to develop the agricultural potential of the state, and he devoted time and energy specifically to apple production in the Bitterroot Valley.[38]

Elrod also helped to found the Montana Academy of Sciences, Arts, and Letters, and delivered the keynote address for the inaugural meeting in December 1902; he served as an officer on numerous occasions.[39] He lectured by invitation in public venues on timely topics such as the potential of Montana forests and the sources of the Montana water supply—"Whence Comes the Water"—for the farmers' institutes funded by the state legislature, for other agricultural and service organizations, and for the teachers' institutes in Kalispell, Butte, and various other towns.[40] He also provided teaching materials to accredited high schools, in 1903 sending out sixty to seventy mounted microscopic slides, adding to some two hundred he had sent in prior years.

More importantly, to his surprise and pleasure, he reported solid results in teaching students with different levels of preparation in the same room, specifically by combining class study with laboratory or fieldwork.[41] Undoubtedly, his success depended in part on the small numbers in his classes, but his dedication to such a unique pedagogical approach deserved a great deal of credit. He boasted that the department's publications matured as well, and that he had initiated exchanges of scholarly work and publications with other institutions. His teaching load included the regular biology courses and a special class on photography for interested students. In 1904, in order to strengthen the collegiate curriculum, he eliminated the prep course on physiology and added one on bacteriology.[42] That year, he also worked with J. P. Rowe, professor of geology and physics, and Eloise Knowles, one of the first two graduates of the university and instructor of art, in preparing the university exhibit for the St. Louis exposition on the Louisiana Purchase. The exhibit included specimens of minerals, flora, and fauna, and a large number of photographs he had taken himself.

Almost annually, in increasingly strident language, Elrod clamored for resources. Even with the largest enrollments in the history of the department, he made certain that every student received individual attention and participated in appropriate laboratory and fieldwork.[43] In accordance with his mantra, every student had "*to learn by doing, to get the information first hand, and to learn to use all of the senses and corre-late the information gained.*"[44] In 1908, he warned the president that the increasing workload threatened to overwhelm one person, a situation that forced him to omit bacteriology that year. Without the appoint-ment of a botanist to help, he predicted departmental collapse.[45] When relief finally came two years later, he exulted that "the division of the work of the department by the organization of a Department of Botany and Forestry has greatly increased the facilities offered to students in biological study, and has made possible more extensive courses of study and higher grade of efficiency in student results. The division was a wise move," he concluded.[46] Repeating his plea for a full school of forestry that required an instructor in botany, he called again for the means to respond to a rapidly growing state industry. Delays continued, much to his exasperation.[47]

Within eight years of opening its doors, the university's growing enrollment reflected the rising number of graduates from sixteen of the twenty-one accredited high schools in the state. The growth imposed new demands for space and courses in physics, chemistry, biology, economics, and literature.[48] President Craig responded as resources permitted, but never quite quickly enough for Elrod and other critics. Nonetheless, the students did well. Several of the 121 university alumni had enrolled for graduate study by 1907 at the University of Chicago, Columbia, Johns Hopkins, Bryn Mawr, Harvard, Stanford, Yale, Dart-mouth, and Michigan, with "not a single case of failure." Even more impressive, two university graduates won prestigious Rhodes Scholarships to study at Oxford University in England: George E. Barnes in 1904 and J. F. Thomas in 1907, and Barnes took a first, one out of eight earned by his cohort of Rhodes Scholars.[49]

Elrod enjoyed the interactions with the students, and they responded accordingly.[50] He also assisted them in organizing a student government, one that persisted in various forms into the twenty-first century, and he served as one of the three faculty members initially elected to that organization.[51] In addition, he acted as the chair of a faculty and student committee that founded the first student publication, initially a monthly that developed into the daily *Kaimin* and thrives today. He thought highly of the student effort and wrote one of the first articles for the new journal.[52] Finally, he collaborated with Professor J. P. Rowe and Coach Hiram B. Conibear to organize and direct the first interscholastic meet at the state university in 1904. This event attracted the participation of twenty high schools and sixty student contestants in track and field events and seventeen schools in declamation and debate.[53] Officially sponsored by the state's High School Athletic Association, the annual meets proved invaluable for recruiting purposes.

These developments manifested the growth and maturation of the university pursued by President Craig and at times denigrated by Elrod and other faculty members. With 30 faculty members, 43 percent of whom had earned doctoral degrees, and 393 students in 1908 spread across 14 departments and 2 schools, an annual budget of $100,000, 2 fraternities and 2 sororities, and 9 student clubs, The University of Montana gradually acquired stature because of its programs, faculty, students, and alumni. However, competition for students with the Agriculture College resulted in lowered standards, in Montana as in other states with distributed campuses, according to the Carnegie Foundation on Education.[54] Elrod continued to support consolidation of the campuses into one university as the most effective remedy, despite Craig's preference.

President Craig attributed the accomplishments to the university's broader outlook and emphasis on practical teaching designed to prepare graduates to deal with the challenges of life in Montana. A sharp focus on the needs of the people and the state, not the aspirations of the institution or the inclinations of a professor, assured that the university responded to all needs within the state: intellectual, professional, and industrial.[55] To that end, in 1898 and again in 1902, he announced the plan to establish a school of pharmacy. However, when the board authorized the school in 1906, Craig declined for lack of resources and space, and the board reassigned the authority to the Agriculture College.[56] Regardless, the president's increasing emphasis on practical, industrial, and professional education struck Elrod as yet another indication of valuing numbers over quality, accepting *more* rather than *higher* education, and reducing the university to the level of a "vocational school."[57] Although he paid little attention to the effects, his penchant to state his opinions frankly, often indiscreetly, combined with his rather dismissive attitude to undermine his relationship with the president.[58] In time, his insistence upon higher standards and adherence to principles rather than accommodating external constraints and seeking compromise led to outright conflict. While Craig's determinedly practical approach served the fledgling university well, it ultimately sapped support for him as an academic administrator.

IV

As time permitted, Elrod aspired to develop the university museum to which he had donated the collection of insects that he had assembled in Illinois.[59] He prized the museum primarily for its usefulness in the education of students, but he also extolled its potential contributions to public understanding of the state's mineral and other resources. He repeatedly—if unsuccessfully—recommended an adequate facility to house the scientific and other collections the museum attracted. When he arrived in 1897, as he remarked, the museum had contained nothing about the natural life of the state, not "a plant, insect, bird or mammal skin." In fact, its small collection had consisted of minerals largely nonexistent in Montana and donated after the Chicago World's Fair in 1893.[60] Elrod boasted that he had built a shell collection representing about sixty-five species of mollusks and land snails in the state, which he kept separate from smaller donations.[61] He envisioned making the university museum a center for the state's natural history, mineral, and other resources for scientific research. However, he made only limited progress toward achieving that goal during his career.[62]

Unfortunately, the museum quickly became a source of irritation because of insufficient space, inadequate support, and careless documentation by both Elrod and Craig. For example, in 1903 President Craig accepted a donation of Indian artifacts from the Missouri River region but the owner insisted years later that he had only intended a loan.[63] At about the same time, Elrod reported receiving a collection of insect specimens, mostly butterflies and moths representing a thousand species, presumably donated by W. C. Wiley of Miles City. In response to Elrod's question about it, the president replied curtly that "it would be well not to pry too closely into the matter, but to consider they were here."[64] Much the same happened with a putative donation of fossil shells by Homer Squyer of Miles City still highly regarded today. Although Elrod recognized the value of the Squyer collection and discredited the others, he nonetheless recommended retention of all the disputed collections without payment or response to inquiries. He doubted that the donors actually had any ongoing interest in the collections, or that they ever intended to claim payment for or return of the donations. While he passionately pursued all collections concerning Montana life, culture, and resources, he exhibited a rather cavalier attitude about the means and details of acquisition and had a dismissive view of the donors.[65]

Elrod's duties increased in 1898 when the U.S. Weather Service placed the former Missoula weather bureau in his charge, where it remained throughout his active career. In 1908, Representative Joseph M. Dixon, at Elrod's request, secured a congressional grant for equipment and student assistants, no small achievement in itself.[66] By that time, however, the weather bureau had lost its luster because Elrod achieved his highest priority when he persuaded the president and the state board to authorize a biological station at Flathead Lake for research and experiential learning. Thereafter referred to as the Biological Station, it became his obsession for the remainder of his career.[67]

Nonetheless, he carelessly placed the station's and his own future in peril because of his casual regard for protocol. As Elrod should have learned by this time, President Craig insisted upon all due respect for presidential prerogatives. When Elrod sent his first annual report on the "Montana Biological Station" directly to the board, he stirred the president to action. Outraged, Craig demanded a resolution from the university committee of the board to change the name to "the University of Montana Biological Station" and to require its director to report progress through the president to the board.[68] Elrod considered the matter trivial, but the president regarded Elrod's action as a deliberate attempt to circumvent his authority. To achieve his goals for the station, Elrod had to exercise administrative acumen and secure adequate financial support, skills he developed only painfully and incrementally over the next two decades. His nonchalant attitude toward detail exacted a significant if unnecessary toll and nearly cost him his position and the station.

Fortunately, he ultimately realized most of his vision for the Biological Station at Flathead Lake. "The object sought in establishing the station," he said later, "was to provide a place where some investigations could be carried on, where kindred spirits might meet to work out plans and ideas, and where students could be taught to collect and study material as it is in the field," with no duplication of university work.[69] In fact, he stated, "Biological Station work consists almost entirely of study of living things within the immediate environment of the station." Elrod defined the major goals as "a better appreciation of living things, plant or animal; another . . . the development of love of nature; and again the student gains in health from the freedom of the life and recreational privileges. Eight hours of work, eight hours of sleep, and eight hours of recreation, including meals, should give health to any one, and supply knowledge of greatest value, either to the individual or to society, or to both." All students at the station had to engage in individual study, often during field trips, occasionally supplemented by lectures, but always including essential materials and competent guidance. In brief, Elrod considered student competency in investigative science as the major object and result of study at the Biological Station. While he welcomed visiting researchers, he always identified students as the primary beneficiaries, and he insisted upon their direct engagement in research focused on the state's natural resources.[70]

Elrod did not invent the concept of a biological station, since other universities across the country had also established them, with Michigan establishing the first one, and California, Massachusetts, Utah, and Washington following suit. However, he strongly believed that Montana had an advantage because of its relatively pristine and variegated environment and ecology.[71] A prominent botanist from Utah, Marcus E.

Jones, who worked at the station with Elrod, assured President Clyde A. Duniway in 1908 that he knew of no other place better suited "for the study of plant life in humid regions," especially the three upper life zones.[72] As he said, "Others have better facilities in certain lines of work, but none approach it in variety." He predicted that supporting a properly designed station at Flathead Lake "would put the University at the head of Biological Station work for students, attracting not only Montana students but investigators from outside in large numbers." While glowingly optimistic about the prospect, Marcus Jones appropriately included caveats about the imperative for support.

The prediction of success ultimately proved correct, although not without a great deal of dedicated effort and frustrating disappointments. Elrod himself experienced only a minor part of the success of the Biological Station at Flathead Lake, but certainly an important part, with a project in 1928–1929 that analyzed the lake scientifically, evaluating its fish population and productive capacity. However, as often occurred during its development, Elrod's station entered a bleak period during the Depression and World War II. The postwar years brought remarkable productivity and the station became a major center of excellence under the direction of Elrod's successors—Gordon Castle, Richard Solberg, John Tibbs, and Jack Stanford. As the "Sentinel on the Lake," one of the oldest and highest-ranked freshwater stations in the world, Elrod's Biological Station at Flathead Lake stands today as a lasting monument to his vision for the future of ecological and limnological research at The University of Montana.[73]

V

Elrod collaborated with Dr. E. V. Wilcox of the Agriculture College to develop the original plan for the station involving a rotating directorship and support from both institutions.[74] For the start-up costs at a still undetermined but suitable site on Flathead Lake, they planned to raise $700–$800 privately. When Wilcox accepted a position in Washington, DC, the Agriculture College withdrew and diverted its funds to other projects, leaving Elrod to raise $750 alone.[75] Future Senator William A. Clark had pledged $250 to the Agriculture College for dredging work at Flathead Lake but agreed to transfer the funds to Elrod when the college withdrew.[76] However, the college resisted because of plans to use the funds for scientific work in the Gallatin Valley. To Elrod's relief, Clark promised another $200 for the new Biological Station at Flathead Lake. With private support, Elrod's dream became reality.

Over the next few years, Elrod located an appropriate site on Flathead Lake and initiated efforts to set up the station. He later claimed that he inspected all of the bays and coves around the lake on the steamboat *Undine*, leased to E. L. Sliter but owned by the Flathead Gun Club.[77] The survey required several weeks in July 1898, camping on the bank of the Swan River a short distance from the inlet to the lake, and he took time to roam the woods and climb the mountains. This experience and a good deal of thought persuaded him to lease land from Sliter on the Swan River inlet to the lake as the most suitable temporary site for six years at minimal cost. In 1908, to showcase the spectacular beauty of the region and to earn needed income, he published a booklet with photographs of the Flathead Lake region.[78] Confident of the attractiveness of the area, he prepared the booklet to educate the public about the majestically beautiful scenery and other attributes of the country. He predicted Flathead Lake was certain to become a popular playground, since almost daily "dozens of pleasure crafts and freight vessels ply the lake's waters." Even with rapid settlement bringing unwelcome change, "the scenery will always remain." Because of the impending influx of population, he considered it urgent to secure a permanent location for the station and to initiate research.

To accomplish that task, Elrod turned once again to Representative Joseph M. Dixon, who had secured the land grant on Mount Sentinel and the funds for the weather station. In the process, however, he assiduously pursued another of his fundamental objectives as well, specifically the preservation of the American buffalo from extinction. By 1902, he had studied the buffalo herd owned by Michel Pablo on the Flathead Indian Reservation, and viewed it as a model for a larger national project.[79] He considered unused reservation land a prime location for a herd, much better than Yellowstone Park with its harsh winters then under experimental use as an animal reserve. He called for an appropriation by Congress to purchase some fifty square miles of reservation land and a small herd of animals, with ongoing management by the Biological Survey of the Department of Agriculture. However, he understood he needed a great deal of assistance to accomplish this goal, so he turned to friends in the conservation-preservation movement.

In 1904 Elrod joined the American Bison Society, founded that year to preserve the American bison (commonly referred to as buffalo) from extinction.[80] President Theodore Roosevelt served as honorary president, bringing along the influence of the Boone and Crockett Club which he, George Bird Grinnell, and others founded in 1887.[81] Governor-General of Canada Earl Grey accepted the role as honorary vice president, and the society included luminaries such as William T. Hornaday (who served as president), Gifford Pinchot, David Starr Jordan (president of Stanford University), H. C. Bumpus (director of the American Museum of Natural History), Madison Grant (secretary of the New York Zoological Society), and T. S. Palmer (Biological Survey, Washington, DC), with Elrod appointed as a member of the advisory committee and subsequently elected to the board of managers.

The society proposed to establish several small herds of buffalo around the United States and Canada, commissioned and paid Elrod "several hundred dollars" to identify a site for a herd in western Montana, and over the next few years discussed possible sites in other states and on the Crow Reservation in eastern Montana.[82] Never a member of the society, George Bird Grinnell worked closely with the members and published numerous supporting articles in *Forest and Stream*.[83] At the society's request, Senator Dixon guided legislation through Congress for 12,800 acres of reservation land—expanded to 20,000—as a permanent national bison range with an appropriation of $30,000 for the land and $10,000 for fences, sheds, and buildings.[84]

Elrod evaluated several possible sites for the bison range, including Wild Horse Island, which he considered too small and too remote from the railroad, and unsuccessful as a herd site during an earlier experiment; the Little Bitterroot Valley to the west near Plains, too remote; the area east of the Flathead River, quite favorable, railroad access, but numerous Indian allotments; and seventeen sections in the Ravalli hills and meadows, the site he recommended because of railroad access and only four or five Indian allotments.[85] He discussed the possible locations with a number of people familiar with the reservation, including Duncan McDonald (the Salish elder), Joseph Allard (son of Charles Allard, who with Michel Pablo had established the Pablo-Allard buffalo herd on the Flathead Reservation), and the Indian agent. All agreed on the Ravalli site because of its nonagricultural but superb gazing land close to the railroad and the town of Ravalli. Elrod extolled the added benefits of a dry and generally moderate climate combined with hills and valleys that afforded protection from the winter winds. However, the proposed range required two and a half miles of fencing to keep the buffalo in and intruders out.

Perhaps of greatest relevance, the Pablo-Allard buffalo herd grazed the area and increased from thirty-six head in 1884 to six hundred by 1907, even with a few sales and some annual harvesting for consumption by Indians and others. Allard had long since sold his interest, and Pablo offered the entire herd for sale when he learned of the decision to open the reservation to settlement. With no other bidders, Pablo accepted the Canadian government's offer to pay $250 per head at the railroad loading site in Ravalli.[86] Elrod shared bitter disappointment with Duncan McDonald when most of the Pablo herd went to Canada. That issue aside, they urged the American Bison Society to place a permanent herd on the Flathead Reservation. As it turned out, some 150 to 200 head of the Pablo herd, mostly old bulls that absolutely defied herding or loading, ran wild on the reservation as late as 1910.[87] Pablo despaired of corralling the beasts and in 1910 announced a hunt-and-shoot opportunity for Canadians to come and kill buffalo for $250 a head, with the heads delivered to the railroad station in Ravalli after the slaughter.

Elrod understood Pablo's original dilemma when he sold the herd to the Canadians, but he considered the hunt unacceptable. After all, Pablo had accumulated a huge profit from an enterprise that had cost him very little, really just the price of the original thirty-six animals plus payment to Allard for his interest, since the animals grazed on reservation grass at no cost for more than two decades. At Elrod's instigation, game warden Henry Avare sought a government ruling to prevent the hunt, and Montana Attorney General Albert J. Galen held that, since the buffalo in question had escaped Pablo's control and resumed their wild state, to hunt or kill them violated Montana law protecting wild animals. When Avare threatened to arrest anyone who participated in the hunt, the Canadians withdrew.

Pablo tried to secure legislation authorizing him to sell the hunting rights to his property, and Elrod resisted that effort as well. The

Bison Society ultimately purchased thirty-four buffalo for $10,000 and received gifts of another thirteen head to stock the new range, which simply accommodated the residue of the Pablo herd. By 1924, the herd had increased to about 700 head, and Elrod urged careful management through harvest and sale to prevent overgrazing and to accommodate the deer, elk, antelope, and Rocky Mountain sheep on the range. Thereafter, he remained an active observer of the range over the years, even as the Biological Station and Glacier National Park commanded his attention.

VI

In response to Elrod's second request, Dixon—who by that time had become a U.S. senator—orchestrated an outright grant that allowed the university to select 160 acres of unclaimed land on the reservation for biological research and education purposes.[88] In 1908, the governor gave President Craig, who in turn delegated to Elrod, the task of selecting the land. Elrod had already discussed possible sites with John K. Rankin, the U.S. allotting agent, and inquired about dividing the 160 acres into various parcels.[89] Rankin cleared partitioning the grant with the secretary of the interior and offered any needed assistance. A handwritten note attached to one of Rankin's letters identified five options: 1) 40 acres at Yellow Bay, about 15 miles south of Bigfork, and 120 acres near Arlee; 2) 40 acres at Yellow Bay, 40 acres on the river at Polson, and 80 acres near Arlee; 3) 40 acres at Yellow Bay, 40 acres adjacent to McDonald Lake in the Mission Mountains, and 80 acres near Arlee; 4) 80 acres at Yellow Bay and 80 acres near Arlee; and 5) 40 acres on Wild Horse Island on the west side of the lake, 40 acres at Yellow Bay, and 80 acres near Arlee.

Elrod stated in 1907 that he initially identified 30.5 acres on McDonald Lake, 53.25 acres at Yellow Bay, 36.25 acres on Wild Horse Island, and 51.5 acres on Bull Island, almost due north across a shallow part of the lake from Polson.[90] However, Rankin informed him that the Department of the Interior had reserved the land around the outlet of the river at Polson and set aside McDonald Lake as a reservoir, and that the Indians had claimed nearly all the land around Arlee. Thereupon, Elrod and Craig narrowed their choices and the secretary of the interior approved the grant of 160 acres in three different locations: roughly 87 acres at

Yellow Bay, slightly more than 37 on Bull Island, and just over 36 on Wild Horse Island. With the land not available until 1910, Elrod persuaded Sliter to allow continued use of the leased site on the Swan River without cost until 1908, thus facilitating a reasonable transition with only a two-year hiatus.

In his station report for 1920, Elrod explained that he had selected the temporary site on the Swan River near Bigfork because of access from the river and proximity to the lake, allowing the steamboat *Klondyke* to dock during high water.[91] Sliter had agreed to a six-year lease for five acres with the stipulation that any buildings constructed on the property were to remain, in lieu of annual rent. At the time, Bigfork consisted solely of two log houses. Elrod used the funds he had raised to construct a small laboratory building (eighteen by twenty-four feet) for $125, and to purchase a sixteen-foot launch for $350 and a rowboat for $35. He chose *Missoula* as the name for the launch, which featured a canopy, seats for eight people, and a gasoline engine. With only the bare essentials, the Biological Station began to function in 1899, although no one received compensation for working there until at least 1912. As it turned out, Sliter did not reclaim the Swan River land until 1909, leaving the station without a permanent site and the equipment in storage until 1912. Nonetheless, the imminent acquisition of the new home buoyed Elrod's excitement and enthusiasm.

He later boasted that he had launched the first gasoline-powered boat on Flathead Lake; it served the station for more than a decade, requiring only a new and more powerful engine.[92] Built in Wisconsin, the *Missoula* came by rail to Ravalli; went by wagon from Ravalli for thirty miles across the reservation to Polson, which then consisted of "a store with a residence attached, a dirt roofed log house, an old log house used temporarily for a blacksmith shop, and a couple of teepees"; and finally by water to Swan River. Only two wooden steamboats traveled the lake during the early years: the old *Klondyke*, later renamed the *Wannigan*, became a mess hall for lumberjacks; and the other, the *Undine*, went through several renovations, sank first during a storm with the loss of one life, and returned as the *Comet* until it sank for good in the Polson harbor.

At the time, people ridiculed the *Missoula*, doubting its utility or trustworthiness. Even the Indian owner of the ferry across the Flathead

River sneered at an engine with "No fire" and "No steam." Because of the importance of water transportation, people demanded reliable and safe vehicles. Originally a three-and-a-half-day buggy drive from Missoula to the station, land transport exhausted travelers.[93] Elrod tried the little known trail along the east shore and found it extremely difficult because of steep hills, wet clay that proved virtually impassable, and fallen trees blocking the way.[94] Water transportation from Polson, while more expensive, offered the best choice before the state completed the east shore highway to Glacier National Park in 1916.[95]

Until after World War II, the station operated only in summers. Beginning with the first session, Elrod launched scientific forays into the mountains and around the lake to collect biological specimens and to characterize the terrain, the ecology, and the lake. Clark's donations in 1900 and thereafter until at least 1910 funded the explorations of the Swan and Mission Mountains. Elrod and his companions climbed various peaks, studied two lakes (Lake Mary Ronan and McDonald Lake), and collected hundreds of bird and plant specimens.[96] In addition, Elrod began to dredge Flathead Lake in order to understand its capacity for fish production.

The students, faculty, and visiting researchers lived in tents and boarded with families in the area until after 1912. Elrod's family visited frequently, sometimes staying for the entire season, and daughter Mary worked as an assistant at the station before and after graduating from the university with a degree in biology. The faculty came from the university and agriculture college or from private colleges and high schools in the state, with a few visiting researchers from Harvard, Chicago, Utah, Illinois, California, Minnesota, and other universities.[97] The station faculty provided personalized instruction for ten to fifteen students; they also conducted explorations and related studies of the flora, fauna, and physical geography of the region. As the "Summer School of Science," the station hosted the university's first summer session.[98]

Elrod and the visiting faculty delivered free lectures in the evenings for faculty, students, and the few community residents who attended.[99] On several occasions, Elrod discussed the geological formation of the lake.[100] After the upward thrusts formed the Mission and Swan Ranges, a massive ice sheet covered the entire Flathead and Mission Valleys and left behind a huge and deep lake that ultimately cut an outlet through the Polson moraine. Until the former glacial lake pierced the moraine, the Flathead River flowed west to the Clark Fork along the course of the Little Blackfoot Creek; and the Swan River flowed north and emptied into the Flathead River near the conjuncture of the North and Middle Forks. As the glacial lake drained through the Polson moraine, the Flathead River took its course to the south, joining the Clark Fork near Paradise, and natural damming forced the Swan River into the lake near Bigfork, leaving the string of lakes toward the north.[101]

The research provided new and valuable information about the summer birds around the lake; the lichens, mosses, butterflies, and botany of western Montana; the forests of the Flathead Valley; the mountains, glaciers, and lakes in the forest preserve that became Glacier National Park; and the mountains from Bigfork to Missoula.[102] More specifically, Elrod pursued his original plan to study the fish and food fish in the numerous lakes and rivers.[103] To facilitate the achievement of that objective—which remained a priority until 1930—he became chairman of the State Fish and Game Commission for a brief period. He also served as chief of the Montana Division of the League of American Sportsmen, and initiated studies of the microscopic life of the lake as fish food. He identified entomostraca (including at least one new species), historically a subclass of crustaceans, as the major source of food for the fish in the lake.[104] In 1908, he made his initial report on the fish and fish food, and in 1913–1914 the Montana Fish and Game Commission provided support to allow him to determine the number, habits, and distribution of fish and any predators threatening the fish in the lakes and rivers. This work involved dredging numerous lakes, including Flathead and ultimately several in Glacier National Park, to establish and identify ways to increase their carrying capacities for fish production. He found many of the lakes in Glacier National Park fragile if not sterile.[105]

Some of the research results baffled him, and he sought expert advice. For example, in 1916 he asked Dr. H. B. Ward, a close friend and an expert on parasites, to identify the tapeworms, some more than a foot long, he extracted from nearly every squawfish he captured.[106] A few bull trout also hosted worms, apparently from eating squawfish. The robust digestive system of the trout transformed the parasite into food and halted the infestation. He also discovered copepods, a microscopic crustacean parasite, and leeches attaching themselves to the gills and fins of bull trout and other fish.[107] Late in his career, after years of studying the lake's capacity for fish culture, he urged an aggressive program to reduce the predators—bull trout, suckers, squawfish, and the alien black bass—by opening the lake for commercial fishing, allowing public sale of bull trout, and putting a bounty on squawfish and suckers.[108] His refusal to tolerate predators conflicted with and overcame his preservation inclination.

CHAPTER TWO

Smelter Smoke, a Painful Ordeal, and a National Park

Despite his engagements with the bison range, the station, and his teaching, Elrod accepted a consulting contract in 1905 to supplement the family income, a decision he came to regret, never mentioned again, and never repeated. His friend and colleague Professor W. D. Harkins of the department of chemistry arranged the contract, and Elrod joined the team of expert witnesses to develop and present evidence in litigation sponsored by the Deer Lodge Valley Farmers Association. The assignment appeared straightforward, and he welcomed the chance to use his expertise for profit. Eschewing a class action suit for damages, the association sought injunctive relief from the allegedly toxic fumes released by the Anaconda Copper Company's new, massive, state-of-the-art Washoe Smelter in Anaconda.[1] To establish federal jurisdiction, the association arranged with Fred J. Bliss, a former Montana resident who had moved out of the state, to file as the plaintiff under an agreement not to sell his land in the Deer Lodge Valley during the court proceedings. Harkins and his mentor, Professor R. D. Swain of Stanford University, conducted chemical testing and analyses of the farm animals allegedly affected by the fumes. As one of the biological experts, Elrod had to demonstrate the toxic effects of the fumes on the vegetation in the valley, presenting specimens and a record of their provenance to validate their authenticity. His reputation as a competent photographer undoubtedly enhanced his attractiveness to the plaintiff team.[2]

After collecting specimens of the affected vegetation in the Deer Lodge Valley in 1905, Elrod testified in June 1906 before a master in chancery charged to hear and assemble the evidence from both sides.[3] Elrod's letters, scholarly papers, and collected newspaper articles contain no mention of this experience. The only direct evidence in the Elrod papers, aside from some notes he prepared after a return to the valley in June 1906, are nineteen photographs he took in 1905 and kept. These photographs include shots of the smelter, some haystacks and fields in the valley, a farmer and a horse near the smelter, some scrub brush in the fields around the smelter, and a hillside dotted with dead trees adjacent to the smelter.[4] The photographs lacked any comparative perspective of before and after to show the alleged effects of the pollution on animals or vegetation, with the possible exception of the dead trees on a bare hillside.

Newspaper coverage at the time ridiculed Elrod's less-than-convincing testimony, and he undoubtedly felt frustrated and embarrassed by the experience, which perhaps explained his reticence about discussing it.[5] He testified that the smelter smoke caused the visually detected damage to the plant specimens he collected in 1905, mounted on cards, carefully identified, and presented in evidence. However, he refused to say whether smoke or disease caused similar injuries to specimens presented in evidence by defense counsel. Without knowing the origins, the conditions when collected, the handling since collection, and all related details, he declined to draw any conclusion. Moreover, he "disclaimed positively any intention of saying that the injuries to vegetation . . . had affected either the yield or quality of the crops," an unusual disclaimer under the circumstances. Further, he "did not pose as an expert on trees, . . . not an expert on soils, not a plant pathologist, not a chemist, not a smoke expert, not an alkali expert and not an authority on hay or crops." Instead, he based his testimony on his general knowledge of biology and the damage over a large area. Claiming only passing awareness of European studies of smelter smoke effects on vegetation, he admitted that he had not read them. As he testified, he relied solely on visual inspections of the specimens for damage, but had conducted no additional analyses of soil, tissue, possible diseases, or climatic changes. Even considering the blatant bias of the reporter in favor of the defendant, Elrod's testimony hardly caused any excitement or received much notice at the time.

He apparently realized the weakness of his testimony and immediately attempted to make amends. Between 19 June and 22 June 1906, he and a colleague returned to the valley, revisited the areas where he had collected specimens in 1905, and collected more for presentation as evidence if requested.[6] He described the smoke effects as "much worse than at any time I visited last year." Even more compelling, he and his colleague personally experienced the "White smoke . . . rolling down over the hill in . . . clouds, several times enveloping us. It was very stifling and irritating to the lungs. Along the roadside the weeds were cooked; dandelions, lupines, and others . . . badly injured." He described in vivid detail the effects "in spots . . . varying much in size from . . . a pin point to . . . the entire portion of a compound leaf. Not a leaf escaped." Even though he found everything smoked, as in 1905, to his astonishment he discovered new leaves on the trees and plants. Overcome by an epiphany, he realized that the new leaves disproved damage by diseases or insects. "They are free from injury and clean in cases where they have grown since the last smoke."

During the hearing in May 1906, Dr. Joseph W. Blankinship, former professor of botany at the Agriculture College who had resigned in 1905 to become a private consultant, had presented the disease or insect theory.[7] Blankinship assembled more than a thousand specimens from as far away as Bozeman to demonstrate the effects of diseases and insects on plant matter. In June 1906, Elrod excitedly recognized that the differences between the old and new leaves disproved Blankinship's arguments. However, that irrefutable bit of evidence no longer mattered because the judge peremptorily dismissed the suit, finding the scientific evidence inconclusive either way and denying injunctive relief to prevent irreparable damage because the plaintiff had rejected the offer to purchase the property at a fair price.[8] Elrod belatedly raised pointed questions about Blankinship's expertise, claiming that Blankinship had no prior experience with damage to vegetation over such a vast area. He also found it amusing that Blankinship actually did not resign from the college but had "lost his position because he could not or would not, or both, carry on the very studies he undertook with the defendants immediately upon his connection with the agricultural college having been severed." These acid comments manifested the resentment and embarrassment of the consulting experience and perhaps vented a desire to get even. In any event, when the contract ended with the dismissal of the suit, he immediately returned to more rewarding activities and never again offered his services as an expert witness for hire.

I

When Elrod eagerly turned his attention to the university again, he noticed improved relations with President Craig, a promising development. In 1906, the president announced that every student admitted to the university after 1 September that year had to have a four-year high school diploma. He also confirmed the preliminary approval of the land grant for the Biological Station and reiterated his plea for adequate and permanent university funding.[9] In addition, he endorsed Professor Harkins's plea for faculty leaves of absence with pay for advanced study and research, and specifically cited Harkins's recent publication on the smelter smoke as evidence of useful university research.[10] That year as well, Elrod's report for the station extolled the work done by twenty-four people during the summer, but advised that the growth of the Bigfork community had rendered the Swan River site unsuitable for further use. He concluded that the destruction of the beauty and utility of the temporary site eliminated any chance of productive research. In any event, Sliter's lease had expired and Elrod longed for a more commodious location.[11]

Elrod's 1906 report also described in elaborate detail an excursion, funded with William A. Clark's annual donation, to Lake McDonald and Sperry Glacier, his first venture after an abortive attempt in 1901 into the area that would become Glacier National Park in 1910.[12] Elrod's party of eight, including Emma and another woman who turned back at Sperry Glacier, went to the glacier, back to Belton, then to Nyack, up Coal Creek, and to Saint Nicholas.[13] With the aims of studying the native tamarack trees of the region and identifying a possible site for the Biological Station, Elrod also made an unsuccessful attempt to climb Saint Nicholas.

Nothing came of the search for a station site in the forest preserve, thus confirming the choice of Flathead Reservation land. The next year, Craig advised the board that the station grant of 160 acres became final with the completion of the Indian allotments in 1910.[14] Elrod's report for the department of biology repeated the familiar complaint of too many students to teach, sixty in total, in too little space, and with only one laboratory assistant.[15] In addition, ten students registered for tutorials at the station, with President Craig as a member of the faculty. Elrod expended what remained of his own time evaluating land for the bison range and the station.

In June 1907, before going to the station, President Craig announced his intention to retire on 1 September 1908, explaining he had agreed to delay retirement at the request of the board. However, a year later he still had made no public comment about his plans after resigning as president, and some evidence suggests that he actively sought a position as president elsewhere.[16] According to Mary Brennan Clapp, the widow of President Charles H. Clapp (1921–1935), who knew Elrod and the campus well, President Craig's final years had become exceedingly difficult because of ill health that exacerbated his sensitivity about affronts to his dignity.[17] In his administrative history, H. G. Merriam simply commented that Craig retired in 1908 because of ill health, well satisfied with the growth and progress of the university.[18]

According to Clapp, Craig and the board discussed his retirement on several occasions, sometimes at his request, before reaching agreement in 1907. Some people suspected then and later that the board actually forced Craig to resign or retire because of personal animosity or alleged concerns about the distorted university focus and lack of institutional integrity.[19] Superintendent of instruction W. E. Harmon, a member of the state board, stated that Craig had become unpopular with the high school principals and teachers, so much so that a majority of the board overruled the state university committee and requested Craig's resignation or retirement in 1907.[20] J. H. T. Ryman, an original member of the local executive committee, also knew the details about Craig's forced retirement and angry response.[21] Despite the public front of an amicable agreement, the unhappy parting of ways made Elrod an easy target.

Setting the process in motion, the board in June 1907 appointed a committee to identify a new president.[22] Although perhaps unknown to Craig or even to Elrod, a number of Elrod's friends sent letters to the search committee nominating Elrod as Craig's successor.[23] A close friend at the University of Chicago, Professor Charles Adams, described Elrod as a sound scholar, great teacher, and man of integrity who had brought recognition to the university. He thought Elrod's familiarity with the educational conditions and needs of the state combined with his accomplishments sufficient to inspire public confidence in him as the next president of the university.[24] Craig's likely response to an Elrod candidacy seems clear enough, although no evidence shows that he knew of it. However, J. H. T Ryman's comment that Craig believed Elrod had instigated the board's request for his departure, combined with Craig's petulant actions in June 1908, suggests he either heard about or suspected the nomination. Harmon, who served on the search committee, knew of the circumstances of Craig's retirement, the Elrod nomination, and Craig's hostility toward Elrod. Elrod never mentioned his interest in the presidency and nothing came of the nomination. Clearly, however, his querulous and frequently tense relationship with the president finally resulted in conflict with potentially severe consequences during the winter and spring of 1908.

In December 1907, President Craig brought the conflict into the open when he charged Elrod before the state board with the misuse of funds donated by former senator Clark for station expenditures during the preceding summer without the knowledge or approval of the president or the board.[25] The charge seems curious on its face, since Elrod had continuously relied on Clark's donations for the station, both for startup and other expenses, including excursions into the Mission Mountains, the Lake McDonald forest preserve, and elsewhere. Moreover, Craig had consistently supported the station and knew of the expenditure of donated funds for construction at the Swan River site and other purposes, such as the trip to Sperry Glacier in 1906.[26] While Craig's claim appears doubtful at best given Elrod's earlier reports, perhaps the president first became aware of the *annual* use of the donated funds for operations and deemed it another deliberate circumvention of his and the board's authority. Whether Craig finally learned of a potential violation of university rules or simply seized the occasion to vent his smoldering agitation, the charge and controversy resulted in a change in the university's financial report for 1910: for the first time, the report listed the Clark funds as budgeted income for approved expenditure.[27] More importantly, the charge of unauthorized use of university funds threatened to end Elrod's career at The University of Montana.

The board discussed the president's charge in December 1907 and again in April 1908, on each occasion instructing Elrod to submit itemized records of 1907 expenditures and to provide detailed accounting for all donated funds in the future.[28] The discussion in April occurred when Elrod, claiming Craig's support, requested up to $400 for station operations in the summer of 1908, since the board had not acted in December 1907, and he thought the June meeting would be too late. He pleaded with the board not to halt operations just as the new station site became available for development.[29] After discussing the benefits of the station to the university, the board approved Elrod's request for funds "to be expended under the authority and direction of the president," and once again ordered "an itemized statement, accompanied by proper vouchers, showing how last year's Clark donation of $250 to the biological station was expended." Either Elrod had yet to account for the use of the funds, or Craig had not accepted and had no intention of accepting Elrod's claim of a mere oversight. Curiously, in view of the allegation, these board actions apparently ended the matter, since it never came up again, and the university released the usual brochure announcing the station's 1908 session with Elrod in charge.[30] As a result, Elrod left the meeting without even a reprimand and with the intent to resume summer work and develop plans for the new station facilities at Yellow Bay.

The board accepted Craig's resignation during the regular meeting in June 1908 and appointed Clyde A. Duniway as the next president, both effective on 1 September. After approving the state university committee report and staff recommendations, the board also adopted the standard motion empowering the state university committee, in concert with the president, to fill vacancies and seek board approval at the next meeting.[31] Until it changed in 1918, university policy limited all faculty appointments to one year, terminating automatically unless renewed. Ominously, on this occasion, President Craig and the state university committee did not recommend Elrod for reappointment as professor of biology and director of the Biological Station. Instead, the board left the position vacant after 31 August 1908, the terminal date for annual appointments. Elrod knew nothing about his impending termination for nearly two weeks after the decision, since he did not attend the meeting. Having secured funds for the summer and having already explained to the president the use of the Clark funds in 1907, he assumed he had no further cause for concern.

Mary Brennan Clapp argued that Elrod's deliberate and unauthorized use of the Clark funds convinced Craig to terminate Elrod's appointment.[32] However, the evidence demonstrates that Craig seized upon a minor oversight, if indeed one occurred, as the last straw in a long-simmering conflict with Elrod, to rid the university of an incorrigible dissident. Upon learning of Craig's rationale, Professor William Aber expostulated scathingly: "I never saw a more obvious case of stealing the livery of Heaven for the service of the Devil."[33] In Aber's disgusted view, the "wrong done Dr. Elrod . . . seems to me so monstrous that I feel about it as Zola did about the Dreyfus affair—to compare a small thing with a greater one." Clapp's argument notwithstanding, Craig's action did in fact reflect his personal animus rather than any concern for his successor or about unauthorized expenditures.

In brief, Craig initiated the action to terminate Elrod's services, and the board approved it without discussion on the recommendation of the state university committee. Putting the decision in context, Professor Aber bluntly explained to President-elect Duniway that President Craig persistently allowed professional differences to become personal. As a result, any professor willing to sacrifice personal interests for the benefit of the university incurred the president's hostile envy. Aber fumed that "The stronger the man the stronger his hatred." Because he openly advocated higher academic standards, Elrod became Craig's mark.[34]

Without doubt, Aber's close friendship with Elrod influenced his views, but other people shared Aber's conclusion that Craig's animus caused the refusal to reappoint Elrod in June 1908. Superintendent W. E. Harmon asked caustically why the person forced to leave had the authority to decide who stayed.[35] J. H. T. Ryman, an original member of the local executive committee, assured President-elect Duniway that Craig maliciously persuaded willing board members of Elrod's disloyalty and untrustworthiness. "He charges Elrod with being instrumental in his retirement when as a matter of fact Elrod was absolutely innocent of any conniving."[36] J. B. Speer, President Craig's and then President Duniway's secretary, identified the outgoing president as the prime mover in the refusal to reappoint Elrod. He also thought Elrod had no chance for a reversal of the decision because good policy militated against reconsidering a decision after the fact.[37] Those closest to the situation knew why President Craig and the board did not renew Elrod's appointment, and the reason had nothing to do with donated funds or concern for Duniway.

H. G. Merriam offered yet another and quite novel explanation of Elrod's near misfortune in a footnote in his *History*, written in 1970 some sixty years later. With no evidence and contrary to all the private discussion of Elrod's predicament at the time, Merriam claimed that Professors Elrod and Harkins "suffered reprisal for testifying for the plaintiffs at the trial of the Anaconda Copper Company occasioned by the harmful effects of fumes from the Anaconda smelter on crops and cattle."[38] For his temerity and at the behest of the company, "Dr. Elrod was dismissed in the spring and reinstated in September at the suggestion of Dr. Duniway." However, Merriam offered no explanation of the reinstatement. As for Harkins, Merriam said the board granted him a leave without pay but "refused tenure . . . and thus an able professor was lost to the University." In fact, the university had no tenure policy in 1908, and none of the faculty had tenure. Moreover, by 1908 Harkins had not attained stature as one of the university's outstanding faculty members, having just earned his doctorate from Stanford in 1908 with a dissertation based on his analytical work in the smelter litigation, which Craig praised. Finally, after the leave without pay in 1908 to complete the requirements for the doctorate, Harkins continued as chairman of the department of chemistry at the university until 1912, when he resigned to accept appointment at the University of Chicago.[39]

No one involved in the Craig-Elrod controversy in 1908 mentioned the smelter fumes case or, after the board meeting in April, the allegedly unauthorized use of donated funds. Merriam assumed that a company-inspired conspiracy explained what happened to Elrod and Harkins in accordance with campus mythology that developed over the years about the Anaconda Copper Company's malevolent influence in the state. The company clearly held considerable sway in Montana, but did not ever control all that happened in the state. In fact, the first mention of company involvement in retaliatory action against Elrod and Harkins appeared in a scurrilous press attack on President George Finlay Simmons in 1936. Merriam discussed the 1936 attack in his chapter on Simmons in the *History,* but did not—in 1936 or 1970—associate it with Elrod's situation in 1908. After Merriam made the charge of company involvement in 1970, Arnon Gutfeld repeated it in *Montana's Agony,* published in 1979, citing an American Federation of Teachers report concerning the Philip O. Keeney case in 1939 that specifically alleged that the Anaconda Copper Company retaliated against Elrod, but did not mention Harkins. None of these accounts offered any evidence, apparently on the assumption that no one needed evidence.

However, as often happens, myth requires only a kernel of truth to attain credibility.[40]

Forty-six years after Merriam leveled the charge and seventy years after its first public appearance, but without citing Merriam or reviewing any evidence, James R. Habeck, a retired University of Montana faculty member, stated matter of factly that the board of education fired Elrod because of his testimony for the plaintiff in the smelter litigation.[41] He later explained privately that he had relied on Merriam's footnote. However, he did not bother to explain why, if Elrod's 1906 testimony angered the Anaconda Copper Company, the alleged retaliation came fully two years later, after the company had prevailed in the litigation. He also argued that several petitions and letters from students and alumni requesting Elrod's reinstatement sufficed to persuade the president-elect and the governor to acquiesce.[42] However, the students and alumni addressed their letters and petitions to the "Honorable President of the University of Montana," President Craig, still in office until 31 August, and it remains unclear whether the president, the president-elect, or the governor ever saw them. Finally, Habeck's concluding contention that Elrod learned from this experience to avoid involvement in controversial issues also missed the mark, as his subsequent career revealed.

II

Developments after the board did not renew Elrod's contract also support the conclusion that Craig sought revenge against Elrod, with the complicity of at least two board members. News of Duniway's appointment reached Missoula in early June. Elrod immediately posted a letter to the president-elect—still at Stanford—extending congratulations and offering to send along plans for the department and the station, when convenient.[43] The terrible spring flood of 1908 isolated Missoula for more than a week and delayed notice to Elrod about the non-reappointment decision until 14 June. Interestingly, over the interim, Elrod delivered the keynote address for a convocation held on campus honoring Craig, in absentia because of the flood, for his long and distinguished service.[44] The president's terse letter to Elrod stated simply that the university no longer required his services after 1 September. To write the letter necessitated awareness of what it meant and why. "I was too thunderstruck to write. I did not get my breath for a week," Elrod exclaimed. Upon reflection, however, he blamed Craig for the decision and identified his own outspoken advocacy for high admission standards and academic quality as the cause.[45]

Without explaining how he knew, Elrod informed the president-elect that most people viewed the decision as a mistake made by less than a majority of the board, and apprised him of Governor Edwin Norris's reputed readiness to convene a special board meeting at Duniway's call to correct the mistake. As for his plans, Elrod stated that he had no intention of resigning, and straightforwardly requested reinstatement to his position. While reinstatement seemed the easiest way to rectify a mistake, he sent Duniway his formal application "for the position of Professor of Biology in the University of Montana, to succeed myself, beginning September first, 1908."[46] For his part, Duniway assured Elrod that he had known nothing of Craig's action until after the board adjourned on 2 June.[47] In fact, he had understood specifically that Craig intended to leave all personnel decisions to him. Appalled and outraged by what had happened, he nonetheless warned Elrod that "you have not helped your cause by some of the methods which you seem to have been following." He obviously had deep concern about a campaign to organize outside influence to interfere with university affairs.[48]

In fact, Elrod's simple and straightforward solution appealed to the president-elect, but he worried about principle, protocol, and board relationships. To an ambitious and punctilious man yet to assume the position he had recently accepted, even the slightest mishandling of this complicated administrative matter threatened to undermine his first presidency. He questioned whether it would be in the university's interest to reappoint someone "summarily dropped from the faculty." In that regard, although difficult for Elrod to accept, Duniway explained that principle, and not personal sympathy, had to prevail.[49]

For Duniway, the principles appropriate to his relations with current faculty members neither required nor precluded his interference with actions of a former president. In his view, the most important principle postulated that faculty appointments depended exclusively "on the initiative and recommendation of the president" to the board without regard to influence. Elrod's reference to the governor and sharing Duniway's letters with A. L. Stone, editor of the *Missoulian,* raised the alarm for a man with an overweening sense of his own rectitude and an abiding respect for propriety and protocol. Since he assumed that Elrod's contract ran to 31 August, Duniway urged Elrod to attend to his work at the station, and promised to decide the matter after a full review. Elrod promptly assured Duniway, "I have been discreet." To avoid any misunderstanding of his stance, he spent the remainder of the summer in isolation at the station, on excursions into the Swan Mountains, or around the lake.[50]

Judge J. M. Evans, the board member from Missoula who said he had deferred to the other two members of the state university committee and approved Craig's recommendation, warned Duniway that reinstating Elrod threatened more harm than good.[51] Either the board retained control or surrendered to faculty and public pressures. Because of Elrod's unspecified but subversive activities, Evans strongly opposed reinstatement, and he thought the other two board members of the state university committee agreed. When queried, board member J. D. Largent stoutly defended the action, arguing it should have occurred long ago.[52] According to Largent, the state university committee considered Elrod disloyal, oblivious to rules, tactless, indiscreet, untruthful, immune to criticism, committed only to his own advancement, and insubordinate to all authority. In any case, as he reminded Duniway, President Craig actually had not made any recommendation. "He simply left blank the Professor of Biology" and a few other positions. What about the few others if the board reinstated Elrod?

Duniway had decided by early July to adopt Elrod's proposed solution but made no public comment until he knew he had the votes to prevail, saying only that he had an open mind on the issue. Nonetheless, he deliberately delayed announcing the vacancy until too late to identify another qualified candidate for appointment in September. Most importantly, he worked hard to persuade board members to support his three-point compromise, outlined in a letter to Evans. As president-elect he considered Elrod the person best qualified for the position. Second, the board had merely deferred filling the position, which unfortunately terminated a good man without charges (and served as a reprimand to the appointee to act with greater discretion in the future). Third and last, the state university committee of four members, including the president, had full authority to fill vacancies and report having done so to the board.[53]

Evans abruptly rejected the compromise since he knew Largent opposed reinstatement and he assumed that G. T. Paul agreed, thus leaving only the president in support. Evans cautioned Duniway that the board never overruled the state university committee.[54] Unbeknownst to Evans, Paul had explained to Duniway that he, Largent, and Evans— although aware of the hostility between Craig and Elrod—had merely accepted the reappointment list Craig submitted, on the assumption that the local executive committee had reviewed and endorsed it.[55] Explaining his position on the issue, Paul agreed to support Duniway if Duniway recommended Elrod's reinstatement. On 5 August, Aber confirmed Paul's support and speculated that "the false impression was a part of a scheme and that the list was made up by Judge Evans with Dr. Craig's knowledge and assent," fully unknown to the local executive committee. In Aber's opinion, "the State Board in full session would *not* have made up a list without Dr. Elrod's name and would *not* have sanctioned such a list save under the impression that the local committee recommended it." Aber vented his outrage at "such plain trace of chicanery," although "*not* greatly surprised."[56]

Duniway also sought the counsel of J. H. T. Ryman, a longtime member of the local executive committee. While that committee had no appointing authority, it played an influential role in university administration until 1909, as Aber's and Paul's comments revealed. Ryman found it amusing but tragic that Evans refused reconsideration, undoubtedly because he realized his error.[57] Speaking precisely to the point, Ryman denied that the local executive committee had ever seen the reappointment list and assured Duniway that no one in Missoula supported Evans in the refusal to consider Elrod's reinstatement. Finally, he assured Duniway that Elrod had nothing to do with any campaign to secure his own reappointment.

Having decided upon his strategy and counted the votes, although still concerned about the possible turmoil if the state university committee divided evenly, Duniway scheduled a meeting of the committee for the week of 17 August, after his arrival in Missoula.[58] To the governor, he extended appreciation for the opportunity to handle the issue and pledged to have it resolved by 1 September.[59] Governor Norris had informed Duniway in late July that he thought the board had erred in not reappointing Elrod, since he knew nothing negative against Elrod except his personal differences with President Craig. Moreover, in view of numerous letters and personal conversations concerning Elrod's reinstatement, the governor pledged to support Duniway even if the state university committee did not agree.[60] Perhaps the governor also received some of the petitions addressed to the president, although he did not mention them. Perhaps Duniway saw them as well, but to what effect remains unknown, since he had already devised a strategy. To board member C. R. Leonard, Duniway confided his willingness to fight for the right results.[61]

Duniway arrived on campus in mid-August and initiated Elrod's reappointment as professor of biology and director of the Biological Station for 1908–1909 with the state university committee, as board procedures authorized. The only record of the committee meeting indicated unanimous approval of the recommendation, dated 24 August 1908.[62] On that day, Duniway and the other members of the state university committee met with the governor in Helena to discuss university affairs. While the governor reported nothing of substance occurred for publication, the meeting clearly resolved the issue.

Duniway's strategy worked because the prior commitments from a majority of the board, including the governor, persuaded the two dissenting members of the university committee to acquiesce in the unanimous recommendation and avoid an embarrassing reversal by the full board. Thus the president-elect shrewdly avoided a divisive contest during the special meeting of the board convened by the governor in September. Duniway's first official board meeting ended in a victory: he won board approval of the university committee recommendation to reappoint Elrod to succeed himself because of the impossibility at that late date of finding a similarly qualified candidate for the fall term. According to the minutes, the board discussed the issue at length before voting unanimously to appoint Elrod with the prior year's salary.[63]

III

Earlier, concerned about his professional future, Elrod wrote Emma from Bigfork of his eagerness for the arrival of the new president.[64] Emma had forwarded a letter from Duniway that Elrod found encouraging. Since Duniway had not requested more information from him,

Elrod anticipated discussion of any issues after the president-elect reached campus.[65] Emma had also forwarded an informative, encouraging, and supportive letter from Aber, one much appreciated by Elrod.[66] Still uncertain of the outcome, he reassured Emma with more confidence than he otherwise exhibited, "Let us hope for the best. If we get the worst we still have each other and Mary, and our reasonably good health." He realized how much depended on Duniway's decision.

To maintain good relations, Elrod kept the president-elect fully informed about his schedule and activities during the summer.[67] With the reservation land unavailable until 1910 at the earliest, the station remained at Bigfork.[68] Elrod used fifteen days to explore the three parcels of land around the lake they had selected, satisfying himself about their best uses. As in past summers, students from across the country came to study in the natural laboratory. Having learned from painful experience, Elrod included a detailed accounting of the use of the Clark funds. The session had proven productive, although time commitments prevented a return to the Lake McDonald forest preserve. Instead, he diverted the Clark funds to a successful trek into the Mission Mountains that produced thousands of specimens for the museum. He also led several excursions around the lake to take soundings and into the surrounding mountains to collect more specimens. Marcus Jones and Joseph Clemens gathered over 10,000 plants representing 1,000 to 1,500 species, and B. T. Butler, a graduate student at Columbia, sought to identify the trees in the dense forests.

Most of Elrod's work focused on the life forms in and around Flathead Lake and the surrounding lakes and streams. He estimated the very diverse and dense flora included at least 2,500 flowering species, all awaiting identification. He observed that western Montana offered the greatest diversity of plant species adapted to cool and damp conditions. The ecology varied from tropical, with huge ferns and orchids, to arctic-like swamps. Many former lakes had long since filled in and become overgrown with new and different vegetation. The forests with unique undergrowth occurred only in the lake region, he claimed. Over the past decade, he and his colleagues had surveyed the Mission and Swan Mountains, finding varied vegetation from the lowlands to the mountains reflecting the differing conditions. They had also discovered rust fungi everywhere, though they had never before observed parasites in such profusion, and collected plant samples with and without the fungi for laboratory study.

He explained the formation of the lake and the region in details he deduced from the environmental evidence. The ice sheet that covered the lake area centuries earlier had attained a depth of 1,500 feet and left numerous "hanging valleys . . . high up in the ravines and gorges" when the ice melted, "mute but indisputable evidence of what actually took place thousands of years ago." But, as he waxed eloquently, "It is only when one ascends the mountains that the grand panorama is unfolded, and the book of nature is spread out, as it were, where an invitation is extended to all who will read." He captured the vista in an exquisitely designed booklet of photographs, that included a multiplate panorama of the Mission Mountains, so that even the new president and others unable to see it personally had the opportunity to read the book of nature.[69]

Daily from 8:30 in the morning until 10:30 in the evening, Elrod labored to identify the life in the lakes, streams, and rivers. He explained, "We want to increase the number of fish in our streams and lakes, for commercial and economical purposes."[70] To that end, the policymakers had to understand the food chain in the waters, extending from microscopic plants and animals to big fish, since the former determined the latter. Unfortunately, the scientific research to establish the limits involved the sheer drudgery of dragging the lake bottoms and lifting the heavy samples into the boat by hand for transfer to the laboratory. The mud drawn from the bottom of Flathead Lake revealed a relatively barren environment, but the shallow areas abounded with life. He specifically identified one species of entomostraca and one blood-red worm at depths of up to 150 feet.

Elrod also visited and began studies of most of the surrounding lakes, including Sylvan, Skag's, Red, and Black Lakes. As far as anyone knew, boats had never before floated on Red or Black Lakes until Elrod launched one with the assistance of F. C. Proctor, a friend who resided at Dayton. Elrod reflected later on a revealing encounter with an Indian passing by as he captured specimens from Black Lake. The Indian stopped his ponies, "startled at the sight of a boat on the bosom of the water. . . . His distant call in impure English of 'What you doin'?' reverberated from the hillside, and the reply 'Catching bugs' created a loud guffaw and a 'giddap' to his little ponies." So much had changed since the invasion of the white man, all to the detriment of the Indian way of life. "Gunpowder has almost exterminated the game, and the modern Indian, a faint imitation of his ancestor, with modern clothing and an Indian wagon jogs along the road nonchalantly smoking the inevitable cigarette and laughing with great glee at the fool bug catcher wasting his time on Stink Lake." Elrod's musings exposed his anxiety about the toll exacted by time and progress and the critical importance of completing his research as quickly as possible.

In his 1908 report for Duniway, a professional historian, Elrod also discussed the pictographs at Angel Point and some other Indian writings he and daughter Mary found on a few rocks near Bull Island.[71] The Bull Island discovery consisted of writings different from those at Angel Point, but out of respect for the integrity of the discovery and his own lack of expertise, he refrained from any comment beyond describing the site in technical detail. He also advised President Duniway of an invitation he received to lecture on the forestry of the Northwest during the Inland Empire Education Association Teachers Institute on 31 August.[72] While he planned to accept, much depended on circumstances beyond his control, with his future still uncertain. He regretted having to delay his return to campus until after 1 September because of the institute lecture, his agreement to harvest tamarack seed for the Royal Botanical Garden in Kew, England, and the need to prepare the station for winter. While as yet a bit unsteady, he had regained much of his equilibrium.

The *Missoulian* reported on 15 September that the board reelected Elrod to the position vacated by Craig's omission of his appointment pending Duniway's decision, putting the best face on the near defeat for Elrod.[73] A. L. Stone, the editor, obviously took pleasure in the report, since he had strongly supported Elrod's reinstatement. News of the outcome elicited a warm response in Missoula, especially on the campus. Stone praised Elrod for his contributions to the university, awarding kudos specifically for the establishment of the Biological Station and the National Bison Range. In the end, Duniway's decision to pursue the right results, even at the expense of a fight, rescued Elrod from his bête noire of becoming an itinerant professor. Despite his narrow escape, Elrod emerged from the shattering experience much strengthened in his commitment to high standards and willing to speak his mind in their defense, although he had finally learned to use more discretion, as President Duniway had counseled.

IV

Duniway's success over Elrod's reappointment before the official beginning of his term heightened his self-confidence and lulled his caution. As a result, within a few months after his arrival in the state, he launched a series of initiatives that sparked more conflict and ultimately shortened his tenure as president. Elrod loyally supported the new president and agreed in principle with most of his proposals.[74]

As a prelude, in December 1908 Duniway urged the board and the legislature to merge all four institutions into one university under a single president governed by a board of regents subordinate to the State Board of Education.[75] He proposed to demote the other three presidents to vice presidents or directors and to eliminate the local executive committees associated with the campuses. Duniway's radical proposal reopened the debate of the early 1890s concerning one or several institutions of higher education in Montana.[76] However, because he doubted the political feasibility of complete consolidation of the four institutions, he advocated administrative unity simply to control program duplication and wasteful competition. As authority for his proposal, he cited statutory language in the acts establishing the Agriculture College, School of Mines, and Normal School that authorized the board to connect them to the The University of Montana as deemed appropriate. He also urged the board to persuade the legislature to adopt the dedicated mill levy suggested as early as 1900 as a permanent source of revenue to support the university.[77] While Elrod strongly supported consolidation, he doubted the merits of Duniway's plan for administrative unity as it finally took shape.

The board responded to Duniway's proposal for unity and the elimination of unnecessary duplication by creating two ad hoc committees, one to consider administrative unity and one to recommend ways to assure institutional cooperation rather than competition and to avoid unnecessary duplication of courses and programs.[78] Misreading the signals, Duniway praised the board action. Reaffirming its intentions, the board in 1911 mandated a timely committee report on duplication, and then clarified its ultimate objective by prohibiting duplication of costly science and engineering programs.[79] Sharpening the perceptual differences about the direction of reform, Governor Norris proposed, and the legislature approved, legislation in 1909 that vested the authority to approve all higher education expenditures in a new board of examiners composed of the governor, secretary of state, and attorney general. The Norris legislation assigned all programmatic authority to the State Board of Education, restricting the discretionary authority exercised earlier by the local executive committees, and also restructured the local executive committees into local executive boards composed of the local president and two members appointed by the governor. The local executive boards had only those functions explicitly delegated to them by the state board.[80] Reform took a direction other than Duniway had envisioned, and he issued a vehement statement castigating Norris's act as an absurd experiment in educational administration that deprived the governing board of the authority to use appropriated funds to achieve institutional priorities.[81] "If an ounce of prevention is worth a pound of cure, it is not too early to begin to consider amendments to a system which might, under some future administration, put the higher education interests of the State in jeopardy through the partisanship of any two of three members of a politically elected board with complete power of the purse."[82] Directly on the mark, his criticism nonetheless elicited a hostile response. The author of the legislation and his appointees and political colleagues dominated the State Board of Education and the Board of Examiners. Nonetheless, Duniway refused to allow this initial setback to subvert his own reform agenda.

Duniway's agenda exemplified progressive higher education reform involving various policy innovations and campus developments. Elrod supported several of the policy changes, including competitive faculty salaries, differentiation of faculty ranks, paid sabbatical leaves for faculty research and advanced study, and higher academic standards within the university. He also endorsed strict requirements for high school accreditation and for admission to the university. However, he questioned the need for the multiplication of professional schools at the expense of the heart of the university—the College of Arts and Sciences—and objected to the reduction of required core courses for graduation, the proliferation of elective courses for students, and the development of narrowly defined major programs of study. The School of Forestry and a School of Law made sense, but Elrod saw little benefit in separately differentiating the proposed Schools of Business Administration, Journalism, Home Economics, Business, Music, and Education. He supported certification to teach for university graduates with appropriate majors who completed required methods courses, and he upheld the strict regulation of fraternities and intercollegiate athletics, but he questioned the effectiveness of an honor code for student conduct enforced only by the students.[83] Finally, he praised the aggressive program for campus development, after completion of the library and heating plant renovation initiated by Craig, but denied the need for residence halls for male students and other student facilities at the expense of academic buildings.[84] Not many of Duniway's initiatives succeeded during his tenure.

The president's demand for action without delay offended several groups of stakeholders. Most traditionalists among the faculty resented the loosening of institutional control over student conduct and the elimination of core academic requirements. Faculty members in the College of Arts and Sciences became increasingly uneasy with the array of emerging professional schools, thinking that they posed serious threats to the traditional curriculum. Numerous high school students, faculty, and administrators complained about the more rigorous standards for high school accreditation and graduation. University administrators and faculty fretted about the impact on enrollment of higher standards for student admission and student work. Administrative colleagues on the other campuses resented Duniway's insistence on the unique status of the university, state policymakers chafed at his aggressive claims for the administrative and management authority of the president, and legislators wearied of his persistent effort to reform the governance of higher education and to increase appropriations.[85]

However, the end came when the president forced an impasse with members of the board by asserting his prerogative to approve all faculty appointments in the new School of Law.[86] Although he won the battle in late summer 1911, he ultimately lost the war. After renewing his contract in June 1911 for another year, the board in December 1911 voted secretly, unceremoniously, and unanimously to allow his contract to expire on 31 August 1912. Duniway attended the December meeting but knew nothing of the board's decision, taken in executive session. Two days after the meeting adjourned, he received the news by mail, much as four years earlier Elrod had, and immediately called a faculty meeting to inform the campus and the community.[87] By giving the facts to the press, he irritated the entire board even more. As he said defiantly, the board "expected me to plead for consideration and to seek humbly for an opportunity to resign." He refused that satisfaction. Instead, he solicited letters from public school leaders and high school students to demonstrate he had support around the state. Personal experience brought into question his counsel to Elrod four years earlier to forego outside influence.

Board member C. H. Hall, the principal figure behind the decision not to renew Duniway's contract, advised the faculty, students, and community that the board had taken a final decision and wanted no debate. While professing to hold Duniway in high regard for his scholarly accomplishments and honesty of purpose, the members unanimously agreed on the nonrenewal as the only way to avoid continued conflict.[88] Some people on and off campus resented Duniway's rigid refusal to compromise on any issue, and still others his misplaced priorities or alleged lack of competence. Eloise Knowles, an art instructor and one of the two initial graduates of the university, stated flatly: "Dr. Duniway is inefficient; he is a millstone about our necks." Professor William Aber wrote that while "good men are opposed to Duniway," he had "not yet heard a good reason from one of them."[89] According to Clapp, even the alumni association refused to support Duniway's reinstatement, although some alumni signed petitions. President David Starr Jordan of Stanford, who had initially recommended him, sent the consoling observation to Duniway that "the reactionaries have taken advantage of their opportunity."[90]

Duniway thought the board refused to renew his contract because of his effort to reform Montana's higher education system. That issue certainly figured in his downfall, but so did his determined insistence upon his executive prerogatives and his rather condescending attitude toward his administrative colleagues on the other campuses. In his final report to the board, excluded as irrelevant from the official minutes and placed in an appendix by board vote, he lamented that "consolidation or even administrative unification of Montana's higher education institutions seems to be a dream, not to be realized because of the strength of the forces of localism."[91] He had sought a public conversation about a critical issue only to find himself banished from Montana higher education. His experience proved prophetic, as several of his successors also suffered the consequences of daring to propose changes to Montana's system of higher education, notably Edwin B. Craighead (1912–1915), Ernest O. Melby (1941–1945), and Carl McFarland (1951–1958).[92]

Elrod belatedly wrote to express deep regret at these developments and apologized that his absence from campus prevented "a hearty official parting grip of the hand." However, he thanked the president for saving his position and career at Montana, "for which I shall always be grateful, and I am truly sorry that our paths diverge."[93] Recognizing that the board had the upper hand and preferred another direction, Elrod kept his peace. In the end, Duniway's earlier advice to use more discretion worked to the departing president's personal disadvantage. Even so, Duniway subsequently enjoyed a very successful administrative career as president of first the University of Wyoming and then Colorado College, and as a professor of history at Carleton College, the University of California, and Stanford University.[94] From afar, he observed with continuing interest the efforts to reform Montana higher education.

V

Edwin B. Craighead accepted the presidency of The University of Montana in 1912. He arrived with the reputation of an academic leader skilled in uniting higher education systems in Missouri, Louisiana, and other states.[95] While accounts differ, Craighead's actions and alliances quickly revealed his intent to consolidate Montana higher education into one institution located in Missoula or Bozeman, preferably the former.[96] He accepted the challenge with robust self-confidence and ambition, combined with a conviction that the agricultural colleges across the country had set out to duplicate the state universities.[97] To prevent them from doing so, he specifically supported federal legislation to hold them to their agricultural and technical missions, hoping to prevent program duplication on a massive scale.[98] Using his significant contacts with progressive Democrats at the national level, he launched the effort to control the agricultural colleges and also sought to secure an appointment as secretary of the interior for outgoing Governor Norris. He ultimately failed to accomplish either goal.

Responding to Craighead's lead, Governor Norris and the board of education resolved in 1912 to support a group of citizens urging the consolidation of three of the four institutions into a "Greater University of Montana," leaving the Normal School with the public schools. Shortly thereafter, on 27 December 1912, the MSTA resolved to cooperate in any way possible with the consolidation movement it had supported two decades earlier.[99] When his bill for consolidation failed in the state senate in early 1913, Craighead led the university in support of a citizens' initiative asking the people to vote in 1914 to consolidate all four institutions on one campus.[100] Elrod and most faculty members supported Craighead and consolidation, anticipating the seat of the new university in Missoula.[101]

Recently elected Governor Sam V. Stewart opposed consolidation, and he quickly appointed a board of education that agreed with him. Praising the quality of the four institutions, he celebrated their strong relations with their host communities, arguing that the state had entered into legally binding contracts with those communities.[102] He also supported and signed into law the Leighton Act of 1913 that authorized the board of education to unify the four campuses into an administratively restructured University of Montana under a chancellor as the chief executive.[103] At his prompting, the board charged a special committee to recommend specific steps to reduce program duplication and implement the Leighton Act. Acting on committee recommendations, the board carried out a series of program transfers. Mechanical engineering went from the university to the Agriculture College; commerce and accounting, journalism, forestry, and pharmacy from the Agriculture College to the university. The board also discontinued several duplicate programs at the Agriculture College, and it also limited the university to secondary and higher education programs and the Normal School to elementary and rural education programs. At the same time, the board formally adopted the Leighton Act, authorized the design of a seal for the restructured University of Montana that now consisted of four campuses, and established a committee on publications to insure that all programs and courses offered by the four campuses reflected their assigned missions.[104] However, Stewart and the board delayed the final restructuring until after the election in 1914 and devoted attention to defeating the public initiative.

As the last step in the process to defeat Craighead, Stewart induced the board to rescind the resolution of 1912 supporting consolidation and to adopt another one strongly condemning agitation and lobbying on campus concerning public issues such as consolidation. In pursuance of that policy, and at Stewart's insistence, the board warned the higher education community against even "the idea of conducting any campaign on any matter pending before the people."[105] The governor and the board members reserved to themselves all involvement in political lobbying, and unsuccessfully instructed administrators, faculty, and staff not to become involved with the public initiative. Ominously, while proclaiming his respect for freedom of opinion, Governor Stewart employed strong language urging all faculty agitators to go elsewhere if they preferred different governance arrangements or thought the Montana institutions inferior.[106] He later stated that the board decided to allow the other campuses to take public positions on consolidation after Craighead and the Missoula campus refused to abide by the board's mandates.[107]

While most state leaders supported Craighead or Stewart, former senator Joseph M. Dixon, also the publisher of the *Missoulian,* grimly warned about the likely outcome of public agitation of the issue. A progressive Republican, Dixon supported the other reform initiative on the ballot in 1914, to authorize woman suffrage in Montana. However, he opposed consolidation as a political pipe dream and detested Craighead, according to some because of disagreements about university printing contracts which Craighead refused to award to the *Missoulian.* Dixon branded Craighead as a dangerous political "pied piper" who deluded people with the "brass band, red fire, and rah rah" style of an inordinately vain man who abused "everybody in Montana who would not pay due homage at his shrine."[108] As a compromise, Dixon urged Governor Stewart to recognize reality and provide funding to "make a Cornell at Bozeman and a Harvard at Missoula." The governor declined, claiming lack of resources, but his actions aptly reflected his personal preferences.

In November 1914, the public initiative for woman suffrage easily won a majority. Not so with the one for consolidation that failed three to two with a 15,000-vote negative margin. To guarantee his desired outcome for the struggle, Governor Stewart vetoed a bill in late spring 1915 to rescind the Leighton Act and subsequently guided the board in restructuring the university under its provisions. In preparation for the new order, the board launched the search for a chancellor to serve as chief executive of the restructured University of Montana.[109] In the aftermath, and once again at Stewart's initiative, the board in June heard a bevy of charges filed against Craighead by J. H. T. Ryman, the longest serving member of the local executive board and Elrod's close friend, including inept administration, drunkenness, fiscal improprieties, and refusal to respect the board's directives.[110]

After tabling the charges without further discussion or explanation, the board simply declined to renew Craighead's contract. Few doubted that his consolidation campaign led to the split decision.[111] Rather than agitate for reinstatement, Craighead accepted an appointment in North Dakota and launched another consolidation effort, which also failed, whereupon he returned to Montana, founded a newspaper to compete with Dixon's *Missoulian,* and died a few years later, some said of apoplexy.[112] The board delayed the search for a new president of the university until after the arrival of the first chancellor in 1916. At the Agriculture College, Hamilton resigned as president because of the stress, but he remained in office until the board appointed his successor a few years later. In 1919, he became dean of men and served for several years in that capacity.

Collateral damage troubled the university for months. At the urging of the university committee, dominated by Craighead's supporters, the board refused to renew the contracts of Dean of Women Mary Stewart and two faculty members, deemed by the public as innocent victims of the consolidation conflict and the futile effort to secure Craighead's reinstatement.[113] They had allegedly incurred the wrath of Craighead's supporters on the university committee for refusing to sign petitions supporting Craighead. Dean Mary Stewart had actually cooperated with Craighead during the initiative campaign, but she also assisted Ryman in the preparation of the charges against Craighead.[114] Curiously, the dean and the two faculty members became the sacrificial peace offerings by the board majority to the Craighead advocates.

At the request of Craighead's supporters and the friends of the dean and the two faculty members, a subcommittee of the AAUP conducted an investigation and condemned the actions against Craighead, the dean, and the two faculty members. However, since the university had no tenure policy, and administrators never earned tenure in any event, the AAUP subcommittee cited only the failure to provide timely notice of nonrenewal of contracts. Even so, the university had no policy requiring notice, and the subcommittee report offered an obiter dictum about freedom of expression, finding no violations of nonexistent university policy.[115] The two faculty members departed and enjoyed successful careers elsewhere, as did Dean Stewart, whom Ryman rewarded with a lifetime annuity after his death in 1926.[116] At the end of the day, Dixon had the last word. He chided a mass protest meeting in Missoula that the agitation over consolidation had ended with the rescinding of the university's charter, making it "now merely a department of the . . . [restructured] university."[117] Duniway's administrative unity had become reality, but some doubted the benefits of that denouement.

Elrod had actively supported Craighead's campaign, but suffered no retaliation. At the president's request, he purchased and became publisher and editor of the *Inter-Mountain Educator* to publicize the issue.[118] Far from remaining on the sidelines and keeping his peace, he fervently advocated approval of woman suffrage and sent form letters to various individuals and organizations soliciting support for consolidation.[119] More specifically, he blasted the Leighton Act as unworkable because it left the four presidents in place and failed to limit their autonomy, despite grouping the campuses under a common umbrella and providing for a single executive.[120] Duniway's administrative unity had little to enforce it, and Elrod scoffed at rhetoric without effect. To inform the public, he published the entire text of the consolidation initiative in the *Inter-Mountain Educator* and urged Montanans generally and teachers specifically to vote for it. Even more dramatically, he warned of the blatant political influence exerted to subvert the initiative effort, specifically identifying the board's rescission of the 1912 resolution endorsing consolidation and the adoption of the resolution prohibiting campus involvement in the campaign at Governor Stewart's bidding.

In agreement, Charles Hemlock reminded Elrod of the rejected deal in 1892 to place the consolidated university in Helena and the state capitol in Butte. He thought the time right to correct the original mistake and adopt consolidation as the only means of assuring a university comparable to those in other states.[121] Elrod blamed the vested interests in

Montana, the public press, the school boards, the other three institutions of higher education, and even the Anaconda Copper Company, for the failure of the consolidation initiative. In the end, he commended the friends of progress for winning approval of woman suffrage, and he applauded getting the consolidation issue to the people for a vote. Despite the outcome, he judged the cause right and certain to succeed "at some future time, when it will cost more. But the more an article costs, the more it is liked."[122] His rhetoric notwithstanding, he never discussed consolidation of the campuses again, and the strength of localism fueled by population growth proved Dixon's prophetic view of consolidation as a pipe dream.

For the first and only time in his career, Elrod had accepted the role of a political activist committed to the consolidation and reform of the state's four higher education institutions into one respectable university. The experience disillusioned him, and he inclined gradually toward the depressing view that time, development, and progress inexorably destroyed all that he valued in life. By the end of the 1920s, he had begun to succumb to reminiscences about, and the ineffable desire to reclaim, the joys and wonder of those early golden years of collaborative and successful accomplishment in Montana.

However, far from avoiding controversy, he maintained an active role in university and public affairs and, as the owner, publisher, and editor of the *Inter-Mountain Educator,* provided a strident voice for education in the state for the next decade. With his university appointment secure, new university leadership offering principled approaches to academic management and programming, the Bison Range fully operational, the Biological Station at Flathead Lake permanently located, and the success of his initial treks into the fascinating region that became Glacier National Park, the prospects for the future of his adopted state and university seemed bright. Elrod eagerly welcomed new opportunities.

The University of Montana in Missoula, viewed from Mount Sentinel.
Four buildings in 1902. Photograph by Morton J. Elrod.

Elrod's butterflies, brought to Missoula with him from Illinois. Basement
of Main Hall. February 1904.

Morton J. and Emma Elrod sitting in an office.

Mary Elrod. Photograph by Morton J. Elrod.

Looking north from the south end of the old bridge, Higgins Avenue.
December 1908. Photograph by Morton J. Elrod.

Catholic mission at St. Ignatius, Montana, in 1899. Photograph by
Morton J. Elrod.

Bigfork, Montana, near the original 1899 site of the Biological Station, taken in 1908. Photograph by Morton J. Elrod.

Flathead River rapids on the Flathead Indian Reservation, Montana, in 1908. Morton J. Elrod.

Five tents and a permanent building in the woods at the Biological Station on the Swan River, July 1907. Photograph by Morton J. Elrod.

Biological Station from boat a few yards from shore at Flathead Lake, Montana, 1913. Photograph by Morton J. Elrod.

Biological Station boat, *Missoula*, pulled up to the shore of Yellow Bay, Flathead Lake, Montana. Photograph by Morton J. Elrod.

Glacier National Park, old dirt road from Belton to Apgar at Lake McDonald. Photograph by Morton J. Elrod.

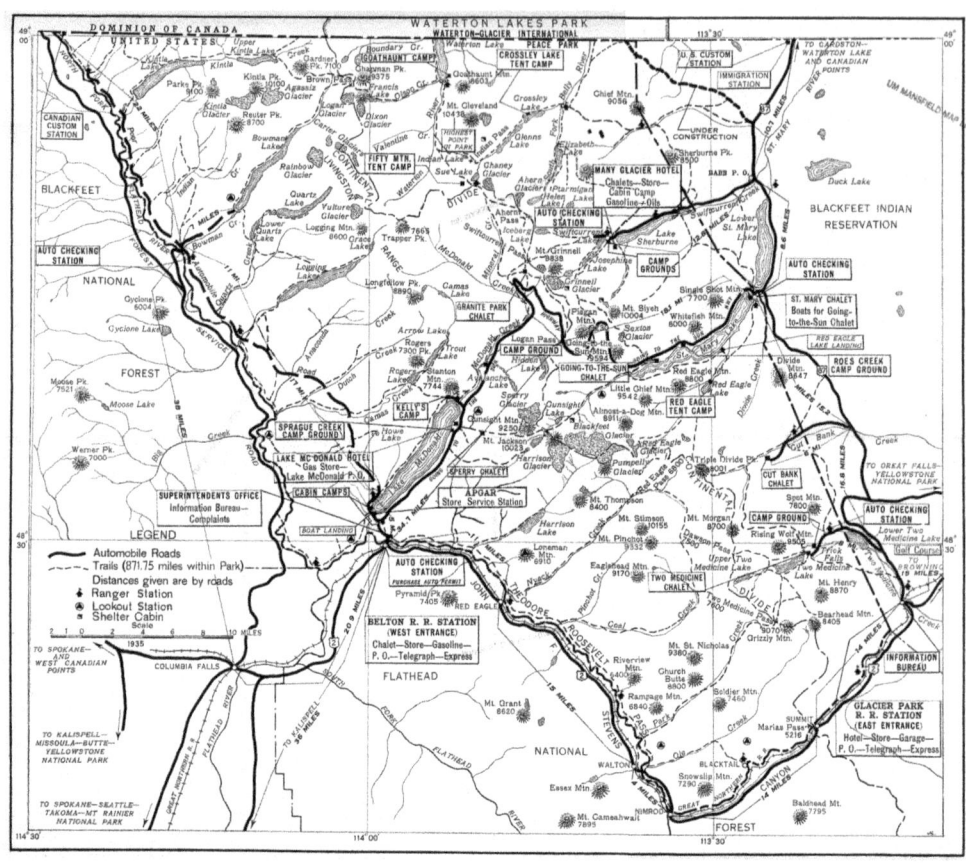

Glacier National Park, 1936.

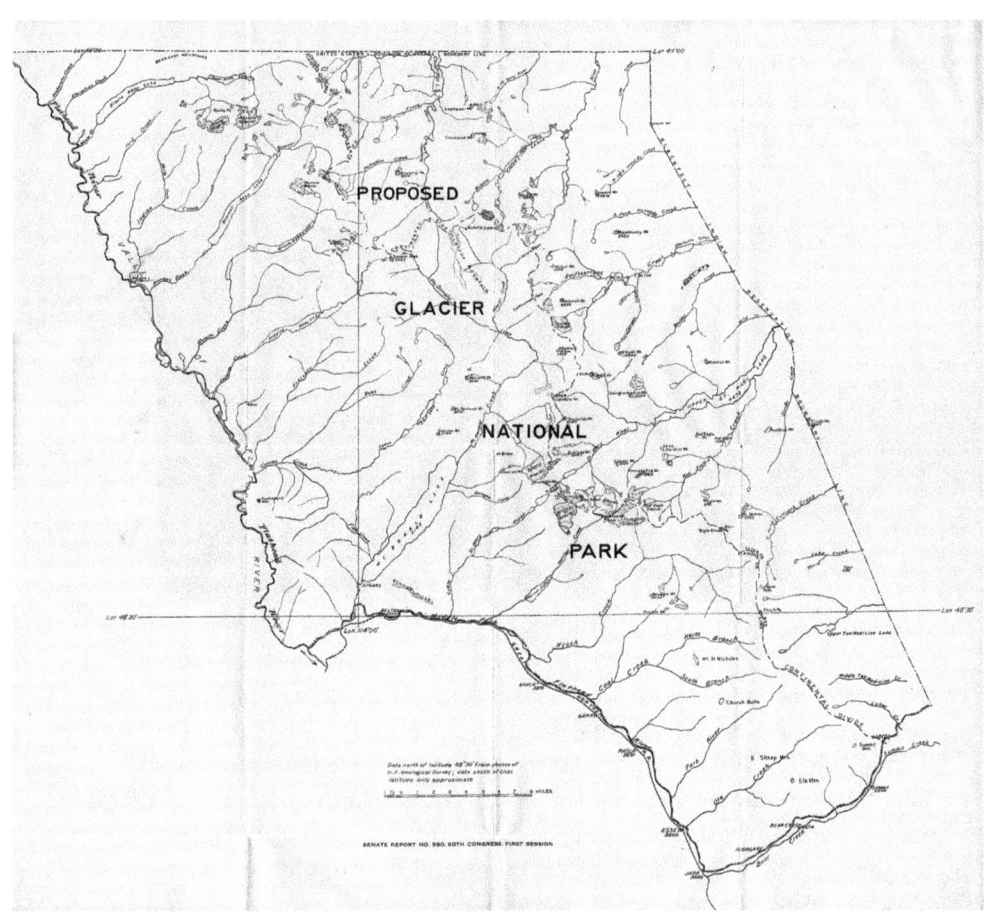

R. H. Chapman map of Glacier Park, 1907–1908. In Senate Report No. 580 by Senator J. M. Dixon.

The lower end of Blackfoot Glacier as it appeared in 1912. Elrod's original caption read: "Blackfeet glacier has a steep descent during the lower portion of its flow. This portion is clean and pure ice, carrying little debris, because the Continental divide at its head has been carried away completely to the ice surface, and the rock material from Jackson and Blackfeet mountains has been left along the lateral moraines. The view here presented was taken in August, 1912. By 1926 the ice had retreated far up the slope here shown. All of the ice behind the moraine, on the right, was also gone." Photograph by Morton J. Elrod.

Elrod with a party at Granite Park, Glacier National Park. Photograph by
E. P. Cole Leamington.

Glacier National Park, Granite Park. 1924. Photograph by Morton J. Elrod.

Horses and riders on a hill above the Many Glacier Hotel in Glacier National Park, Montana, circa 1920. Photograph by Morton J. Elrod.

Morton Elrod, Winold Reiss, and George Bird Grinnell on Grinnell
Glacier, 1926.

Glacier National Park, construction of Going-to-the-Sun Road.
Photograph by Morton J. Elrod.

Morton and Emma Elrod in Glacier National Park, Montana.

CHAPTER THREE

A New University

Over the next two decades, Elrod immersed himself energetically in university governance and his academic engagements even as he passionately pursued his obsessions. The university's perennial struggle for funding and public support became more difficult because of the turbulence of World War I, the postwar recession, and the Great Depression. During this lean era, Elrod fulfilled his campus duties and helped to guide the university through its transition from the formative years to maturity. Given the worsening economic conditions, he understood that his vision for the Biological Station depended on aggressive leadership to muster the necessary resources. His struggle to assure his vision ultimately ended in a conflict that delayed the station's full realization for two decades. Having made successful excursions into Lake McDonald country, he determined to see and learn more, driven by a moral imperative to preserve the pristine mountain wilderness as a national park and public playground.[1] He had gained insight about academic politics from his ordeal in 1908 and the consolidation campaign. However, he soon found much to dislike about the recent trends in higher education. Predictably, he spoke out about the issues and worked hard to find solutions to all of them. The volatile economic and social changes of these years threatened to overwhelm even the best-prepared participants.

As happened elsewhere, Montanans reacted to World War I initially by preferring American neutrality.[2] However, when President Woodrow Wilson found war necessary to protect the rights of neutrals, Americans including Montanans by and large acquiesced. In fact, Montanans volunteered eagerly for military service, and public opinion rallied in support of the war effort, even for the legislation that branded public criticism of the war and war measures as sedition.[3] In Montana, the state legislature pioneered the statute against sedition, finding the national Espionage Act insufficient to halt acts interfering with the war effort and subverting the government, and established the Montana State Council of Defense.[4] The sedition legislation served as the model for the federal Sedition Act of 1918, and the council quashed public agitation against the war. It also prohibited teaching German in the high schools and on the campuses of the restructured University of Montana, banned the sale of works in German, and required the removal of books in German or about Germany from library shelves.[5]

Economically, the government war efforts furnished strong stimuli for agriculture and manufacturing, all closely regulated by administrative fiat, creating auspicious precedents for the future.[6] On the other hand, funding for domestic programs suffered, especially for education, and the war demand for human fodder dramatically reduced college enrollment. Life in general changed radically during the war years, as the government strove to maximize production of food, war supplies, and munitions; willing and able individuals had little difficulty finding gainful employment. The advent of the automobile and highways to promote the transportation of war materials and personnel introduced a new mobility and temporary prosperity on the home front that further loosened any remaining ties to the old village society of the nineteenth century.

At the State University, the university's new name after restructuring, the faculty protested the stringent restrictions on academic freedom, especially those against teaching German language, literature, and history.[7] President E. O. Sisson issued an eloquent statement in defense of free expression and academic freedom, and he refused to participate in the activities of the local Defense Society (not affiliated with the state Council of Defense) in Missoula because of its threats to constitutional rights.[8] Despite Sisson's public stance, university faculty, staff, and students shared the propensity to view criticism of the war, war effort, and political leaders as unacceptably dangerous. Several university programs were temporarily suspended for want of faculty and students, with war needs taking priority.

To attract and retain students, the State University secured approval of a federally funded Student Army Training Corps (SATC) that provided clothing, supplies, and tuition for the cadets who attended special courses taught by university faculty members.[9] However, the war ended before mobilization of the SATC, though several students enlisted, and some made the ultimate sacrifice. To honor the students who died in the war, Elrod and Dean A. L. Stone proposed and supervised the planting of evergreen trees along former John Street, from the northern boundary of the campus to University Hall.[10] In addition, a faculty committee unsuccessfully proposed as a monument to the fallen Montana students a Greek cupula in the center of the university's oval, with openings to walkways dividing the oval into quadrants.[11]

During the war, the state froze or reduced all funds for salaries, for repair or renovation and construction of new facilities, and for new initiatives of any kind. Higher education fell to the bottom of state priorities, with the war effort overwhelming everything else. Even before the peace treaty that marked the end of the war, Montana and the country had plunged into a period of social turmoil fueled by perceptions of radical socialist or communist tendencies unleashed by the war, the Russian Revolution, and a rash of labor strikes for more equitable pay.[12] These distractions continued and led to frantic efforts to assure stability that hindered postwar stabilization in the country.[13]

Despite the constraints, new chancellor Edward C. Elliott persuaded the state board to request proposals from the campuses for everything from new academic programs to meet postwar needs to staffing patterns, salary requirements, and comprehensive facility master plans. Elrod served on the committee that developed an ambitious facilities plan for the State University including a new library, forestry building, and science building, all as part of the new campus master plan developed by Cass Gilbert at the chancellor's request.[14]

To finance these developments, the chancellor proposed two public initiatives in 1919, one to dedicate the revenue from one and a half mills for the operational expenses of the University of Montana and the other for a bond issue of $5 million for campus construction.[15] The board approved the initiative campaign on the condition that the chancellor, presidents, and alumni associations had to conduct it. In the election of 1920, the public finally responded to the persistent requests for reform of higher education financing and approved the two initiatives. In his

Charter Day address for 1921 at the State University, Chancellor Elliott eloquently welcomed the opening of a new era with adequate funding and facilities.[16] However, even before the first dollar was spent, the economy and public mood changed and hard times returned to plague higher education.

I

By 1929, the State University, one of the four campuses of the restructured University of Montana, had become a comprehensive undergraduate institution. The faculty and administration affirmed its maturation in a petition for a Phi Beta Kappa chapter that year, although the effort failed because of budgetary, curricular, and faculty inadequacies.[17] During the difficult 1920s, economic necessity and individual choice had pushed and pulled students away from the basic foundation of a liberal arts education and toward professional or vocational training. Such educational choices increasingly linked coursework to workplace requirements and allowed students to bypass essential grounding in language, mathematics, science, and culture. In the eyes of critics, quality gave way to quantity, rigor to expediency.

In this changing scene, Elrod indulged fond memories of a time marked by "a spirit of study for the purpose of acquiring knowledge and becoming acquainted with the world's best things in science and literature, rather than pursuing such subjects and such only as may be directed to money making channels. If we have not gone mad on the utilitarian studies we are certainly strongly headed that way."[18] He and other dissidents thought informed choice possible only when exercised after consultation with wise faculty advisers.[19] These early critics anticipated Harry Lewis's strictures almost eighty years later decrying the advent of "Excellence Without a Soul."[20] Elrod and his colleagues at times seemed to view educational reform and progress as illusions. In their eyes, progress, like a goose, consumed everything before it and befouled all behind it, the lament of generations of American conservatives. However, Elrod's insistence upon indisputable evidence as the foundation for judgment in the end sustained his faith in progress.

Although he approved its celebratory tone, Elrod took no direct part in the Phi Beta Kappa petition of 1929. Academically, he supported teacher preparation programs, protested the proliferation of professional

schools—such as schools of education and library science—unless staffed
with arts and sciences faculty, and strove to attract more students to the
sciences. His recommendations for curricular changes and demands for
support became less strident every year for a variety of reasons. Perhaps
he received more of the assistance he needed. Possibly, advancing age or
his summertime escapes to the Biological Station and Glacier National
Park mellowed him. Quite plausibly, he finally recognized discretion as
the better part of valor, or he found President Clapp more congenial,
responsive, and attentive to his concerns. Be that as it may, his official
duties became time-consuming when the chancellor appointed him vice
president of the State University for one year, with duties the president
delegated.[21] Despite its growing complexity and the expertise of Chancellor Elliott, the University of Montana did not implement a modern
administrative structure on its campuses until after World War II.

Adding to his responsibilities, from 1925 to 1930 Elrod served as
chair of the Inland Empire Education Association Teachers Committee
to study and recommend curricula for science education in the public
schools of the Northwest.[22] He took direct responsibility for the committee's recommendations concerning general science, by far the most
popular course, which he thought should be taught in the grade schools,
not the high schools. Predictably, he emphasized mandatory field and
laboratory experiences as the essential foundation, with textbooks only
for review of facts, leaving theory for high school and college. His success in this endeavor led to his election as president of the Inland Empire
Education Association for 1929–1930. His presidential address in 1930
reviewed the challenges confronting the schools and urged teachers to
fulfill their societal function in the promotion of acceptable change.[23]
Perhaps poignantly, in view of its repeal in the 1930s, he credited the
teachers in the "the little red school houses" across the country whose
pupils as adults voted overwhelmingly for prohibition to the benefit of
the society. Fulfilling his academic assignments while attending to these
obligations exacted a toll in time and energy.

II

In an overview of American higher education from the beginning to the
early twentieth century, Elrod condemned the increasing emphasis on

extending university programs and courses to the surrounding communities for demeaning and cheapening a noble endeavor.[24] This misguided effort opened higher education to all comers regardless of preparation or motivation and resulted in lowered standards to accommodate everyone. He estimated that the United States enrolled one person in college for every four hundred citizens, or every three hundred including the normal schools, a ratio far too high for effective education. The flood of students inundated the faculty. To manage the numbers, temporary and part-time instructors did most of the teaching, with fewer full professors and a rising percentage of instructors and assistants.

As have modern critics, Elrod attributed the deteriorating situation to uncontrolled student numbers, stagnant or declining resources, and rampant competition for ever more students. Inevitably, too few students came academically prepared and fewer still ever actually interacted with accomplished scientists and scholars.[25] Regrettably, the faculty ranks attracted only lesser lights because of better rewards and benefits in the trades or professions. He argued, as he had during the dispute with President Craig years earlier, that the substitution of the high school diploma for the entrance examination started the decline which culminated in the segregation of courses into groups of so-called electives for student choice, the proliferation of narrowly defined academic majors, and the omnipresent professional school. "Educational institutions . . . have gone mad on the utilitarian side, due largely to the development of the professional school."[26] Rather than a broad education for life, students chose what to study and invariably followed the line of least resistance. In Elrod's opinion, only standardized admission examinations administered by an external agency and a structured education designed to meet individual needs had the potential to assure human progress.

Nonetheless, his analysis of developments since about 1870 convinced him of beneficial changes almost beyond belief.[27] He speculated that the ingenious creation of history as the source of societal memory enabled succeeding generations to extend the frontiers of knowledge. "Any talent arising at any period adds decidedly to the knowledge which may become the possession of those less talented." Unfortunately, two historic and seemingly innate human weaknesses limited and perhaps threatened to end progress: 1) the powerful rule of tradition in human culture; and 2) the rarity of creative insight. In Elrod's view, these human failings resulted in rising numbers of less endowed people seemingly incapable of benefitting from education, thus threatening even the progress engendered by previous generations. Elrod's meditation about the rise of the less endowed or poorly situated groups emphasized the peril to society.

As evidence for his argument, Elrod adduced the facts "daily brought to our attention . . . of the insane, the criminals, the feebleminded, the paupers, and the indigent." This horde of dysgenics already exceeded a million in the total population of 120 million. He thought one salient fact illuminated the impact: "Montana pays more for the support of this group than for her institutions of higher learning." Ominously, "Insanity and crime are increasing. The oriental problem is no small problem. The negro question has not been settled." Urban decay threatened to engulf a society increasingly dominated by large corporations. Nor had anyone demonstrated definitively how to measure human ability, and educators had yet to invent a responsive curriculum to educate people for social order and the appropriate use of increasing leisure. Of critical importance, a stable society required laws and programs to protect the family and the institution of marriage, all regrettably lacking.

In a piece Elrod initally presented to the Cosmos Club and subsequently published in the *Missoulian,* he suggested the possible use of eugenics to rid the society of the "mentally unfit and the criminal class" through mental and physical testing and an appropriate free education.[28] Attracted by the spirit that infused Hermann Muller's "Geneticist's Manifesto" (not published until 1939) to improve the human species, Elrod recognized the dangerous implications even before the Nazis demonstrated the consequences of the abuse of genetics during the years after 1933.[29] In fact, he stopped well short of the measures implied by some of his rhetoric. Since the emerging science of genetics had not established the laws of heredity capable of preventing physical and mental disability, Elrod invoked—perhaps futilely if idealistically—a structured education, particularly science education, as the means to surmount the challenge.

As a beginning point, he advocated mandatory, standardized, and periodic mental and physical examinations to establish the potential of each student and to teach each properly. In his view, "No one has a right to ask or expect more, and the discovery of ability must be possible, no matter what may be the objection or protest." Logic revealed the fallacy of uninformed elective study, dangerous to students and society alike. He also brought to bear the results of scientific research and common sense that illuminated "the folly of uniform and standard work for each and all." Teachers had to tailor education for the needs of each individual and must never ignore the superior student, in Elrod's opinion the only source of creativity in human affairs.[30] Most important, he predicted the approaching insufficiency of resources to cover the costs "of caring for the dysgenic classes" without a societal effort to limit the numbers and protect civilized society.

In the end, science and learning by doing had the potential to achieve that goal by fostering creativity and by neutralizing tradition with the corrosive power of fact and logic. However, the education of scientists and science-minded people required direct and personal engagement with the natural world. In short, science-minded people had *"to learn by doing, to get the information first hand, and to learn to use all of the senses and correlate the information gained."*[31] Science-based education empowered people to challenge authority, to test the validity of practices and traditions, and they willingly brought this valuable asset to the service of society at large. Elrod specifically advocated educational strategies capable of inspiring "enlightenment" rather than rote or routine learning. By enlightenment, he meant the motive power "to shape the human race," an insight he borrowed from the celebrated neurologist and educational reformer Dr. Joseph Collins.[32] Collins had argued that "When the fundamental principles of biology, physiology, psychology, and sociology are taught in the schools . . . at the expense of algebra, history, and rhetoric, race improvement and individual happiness will gain momentum." Collins's ideal curriculum satisfied Elrod's vision but stood little chance of acceptance in the twentieth century.

III

Despite the governance changes at the State University, the president continued to dominate internal policy under the oversight of two rather intrusive chancellors. Edward C. Elliott, the first chancellor, came to Montana from Wisconsin with a reputation for pioneering the use of statistical data in educational administration; as a founding member of the AAUP and its Committee on Academic Freedom he had advocated for academic reform.[33] He accepted the position only with a board commitment to implement faculty tenure as well as new processes and procedures for employee suspensions and terminations.[34] After his arrival, he promoted faculty involvement in the shared governance of the campuses, within administratively imposed limits.[35] Even the relatively new President Charles H. Clapp observed in 1922 that Chancellor Elliott's intrusive interference in minor details undermined the positive effects of his reforms.[36] After more experience, however, Clapp came to view the chancellor's coordinating function as critical to the success of the restructured university. It encouraged respect and cooperation among the four different schools, each vying for attention and funding.[37]

Reflecting on his experience with the consolidation campaign, Elrod sought to effect change by advising the new chancellor. As a crucial first step, he urged the chancellor to define the faculty as the voting employees and to delineate their responsibilities and prerogatives.[38] Under current arrangements, too much university business occurred without faculty involvement, an absurd arrangement for an academic institution. He also advised eliminating the now superfluous presidents and instituting elected chairs of the campus faculties with no executive authority, since the chancellor had become the chief executive officer of the restructured university. His third proposal called for drastic reduction or complete elimination of deans, the source of discontent on the campuses. Effective academic governance required controlled management of departments and programs to prevent the professional schools from interfering in general university business beyond their authority. In his view, the schools had become virtually autonomous to the detriment of sound university functioning. As his final caveat, he warned against more dormitories on the campuses as catering to numbers rather than standards. Students easily found housing in the community, but only the campus offered education. Elliott and the board subordinated the presidents in the interests of unity, cooperation, and efficiency, and initiated new policies recognizing a broadened faculty role in institutional governance. However, Elrod's disdain for deans and student dormitories gained no traction.[39]

Upon his appointment in 1916, Chancellor Elliot led the board in implementing several critical policy changes.[40] The chaos left in the wake of failure of consolidation surely reinforced the need for reform.[41] Standing aloof from the board's suspensions that ended the consolidation crisis, Elliott sharply criticized the AAUP investigating subcommittee for delaying his efforts to develop and implement the much-needed policies. He thought the subcommittee ignored its charge rather than acting to prevent similar problems in the future.[42] The entire episode struck Elrod as all too familiar because of his own experience in 1908, and he welcomed the chancellor's reforms.

Within two years, the restructured University of Montana had instituted shared governance on the campuses. Progressive policies and procedures ensured timely notice of suspensions or terminations of appointment, with hearings if requested, and established faculty tenure after three years of satisfactory service, terminated thereafter only for cause following a hearing or because of age.[43] To hear complaints about suspensions or terminations, the new policy mandated a committee on service to consist of three faculty members on each campus, one appointed by the faculty, one by the president, and one by the chancellor. Although Elliott formed a faculty committee for advice and counsel about the policies and consulted the presidents, the advice made little difference. Elrod served as the chairman of a state university committee that submitted several proposed amendments to clarify certain policy issues, but nothing changed. Reflecting on the experience, he commented waspishly that the faculty actually had no role in developing the policy.[44]

More to his liking, consultation with the faculty became the standard approach on the State University campus. While the faculty had met in response to the president's call to approve curricular changes almost since the founding, faculty advice on policies had rarely made much difference. Change began in 1918 when President E. O. Sisson sparked controversy by reducing the leave pay of Professor William Aber, a member of the original university faculty, without consulting anyone.[45] After discussion with Chancellor Elliott, Sisson proposed a faculty welfare committee, with Elrod as one of the original members, to advise the president on administrative and personnel matters.[46] Elliott and Sisson viewed the committee as the first step toward effective shared governance of the university. Elrod also became the first chair of the state university committee on service, and held the position continuously until 1934.[47] In his capacity as chair of that committee, he issued interpretations of the policies and procedures that guarded faculty rights: the freedom to teach, conduct research and publish the results, provide campus and community service, and exercise the rights the U.S. Constitution guaranteed to all citizens. Elrod's fervent commitment to these rights informed the new policies and procedures when implemented.

In February 1919, Chancellor Elliott suspended Professor Louis Levine, a tenured economics department professor, without pay for publishing a report on mine taxation that the chancellor had initially requested and provided assistance for Levine to produce as a university bulletin.[48] When Levine clashed with friends of the Anaconda Copper Company in public meetings, the state senate received complaints about rampant socialism at the State University, and Governor Stewart demanded a full report. In view of these developments, the chancellor had second thoughts about the project and ordered Levine not to publish under any circumstances. If the Anaconda Copper Company exerted influence to assure this development, however, the effort aborted.[49] The state senate investigated and found no evidence of socialism at the university, and Levine secured a publisher on his own but the book stirred only minor discussion in the state, as equitable mine taxation had gained substantial support.[50] The board and President Sisson called on the State University's committee on service to investigate the suspension and present its findings.

Elrod's committee straightaway denied the authority of the chancellor or the board to interfere with a faculty member's right to publish the results of legitimate research and stand accountable in the marketplace of ideas. According to Elrod's draft of the report, softened somewhat in the final version, a "faculty member is more than a hired man" in his institution and field of expertise. By analogy, a faculty member had the "same relation to the Executive or governing board as does a judge to the governor or president who appoints him." Appointed because of "special fitness" for his position and work, he must have "the utmost freedom in the discussion" of his field of expertise with students and the public. In brief, "He alone should be responsible for his utterances, and not the person or persons who appointed him. Outside of the class he should have the same freedom as other individuals . . . to write or to publish, and should be encouraged to present investigations in his field of endeavor, for by such methods only will progress be made." The committee on service noted and agreed with President Abbott Lawrence Lowell of Harvard University that no middle ground existed: if faculty members did not stand accountable for their speech and publications, then the university incurred the responsibility for all that they said or wrote.[51]

Equally revealing of his stance, Elrod discussed the case fully in the *Inter-Mountain Educator* despite his role as chairman of the committee, and he included details available only to the committee.[52] He obviously saw no conflict of interest in reporting news of benefit to teachers, whatever his part in the proceedings. In the end, the board voted six to three to sustain the chancellor's suspension and seven to two to reinstate Levine with back pay.[53] Mary Brennan Clapp commended a decision that simultaneously supported the chancellor, defended the faculty member's rights, and respected Elrod's committee report and recommendation.[54] President Sisson, who regretted the chancellor's suspension of Levine, welcomed the final decision as a triumph of principle.[55]

However, Elrod understood the outcome quite differently, and took the board of education to task for upholding the chancellor's suspension yet inconsistently reinstating Levine with retroactive pay to the date of the suspension, without affirming Levine's right to publish.[56] Clapp put the best face possible on the incident, arguing that Elliott suspended Levine to avoid his suspension by the governor (acting in his capacity as chair of the state board), thus allowing time for the dust to settle. She suggested that Elliott understood that suspension by Governor Stewart at the board's request all but guaranteed dismissal. The AAUP investigator took a much less nuanced view and concluded without doubt that Elliott suspended Levine for insubordination.[57] Nonetheless, fully restored to his position and with no financial loss, Levine resigned to accept a position with the *New York World*.[58] Elrod insisted thereafter that the Levine report and board decision should settle all such cases at the State University, as did the AAUP Committee on Academic Freedom.[59] Clapp's interpretation notwithstanding, Elrod's assessment hit the mark.

The second case involved the technical issue of reassignment of a faculty member serving on a term contract rather than suspension or dismissal.[60] Scion of a wealthy Chicago family and educated at Harvard,

Arthur Fisher came to teach in the State University School of Law in 1920 on a two-year term contract. Almost overnight, he alienated several powerful citizens—including the editor and publisher of the *Missoulian,* the executive committee of the Montana American Legion, the members of the Montana Newspaper Association, and others—when he became part owner of a community newspaper. These citizens became most concerned when the community newspaper espoused reform and advocated ideas they considered radical, such as the MSTA becoming affiliated with the American Federation of Labor (AFL).[61] That Fisher had sought and secured exemption from the draft during World War I because of a disability—some alleged as a ruse—and then participated in protests against President Woodrow Wilson's refusal to state his war aims did not escape notice.[62] After a star-chamber interrogation of Fisher in 1921, the Legion's executive committee demanded that the board discharge him from the faculty because of his radical ideas and malignant influence on young people. Fisher, the board, the chancellor, and the president referred the charges to Elrod's committee on service.

Elrod's committee found no reason to interfere with Fisher's contract, since he had not violated law or policy. The administrators expressed a somewhat different perspective. While they agreed to dismiss out of hand the Legion charges against Fisher because they focused on matters of personal opinion, the chancellor, president, and dean disagreed with the committee about the quality of Fisher's teaching. Elrod's committee considered the teaching at least satisfactory, while the chancellor, president, and dean insisted that Fisher needed time and effort to improve his teaching, and that he devoted too much time to partisan activities with the community newspaper, neglecting his university assignments.[63] As a result, the administrators recommended probation for Fisher, with a mandate to improve his teaching and attend to his university duties. They also urged the board to delay any decision about his contract until April 1922, the policy date for notices of nonrenewal.

For the committee, Elrod held that the Levine precedent applied to Fisher's outside activities. Yet after dismissing the Legion allegations as irrelevant, in obiter dictum the committee report rejected the idea that faculty members had to avoid "all subjects of a controversial nature" and "teach only undisputed facts." As even the AAUP subcommittee conceded, academic freedom did not appear as an issue in the case.

Agreeing that faculty members must exercise "common sense and good judgment," Elrod's committee categorically denied that anyone had the authority to tell faculty members when to speak or what to say. However, the denial came as a gratuitous remark. For its part, the board majority dodged all issues by placing Fisher on leave with pay for the remaining year of his contract and providing advance notice of nonrenewal of his contract on 31 August 1922.[64] Two years after the fact, an AAUP investigating subcommittee—in obiter dictum because Fisher's term contract had lapsed, thus mooting the case—found the board had committed a severe injustice by preventing Fisher from teaching, in effect disallowing his professional service while nonetheless paying him.[65] However, neither the AAUP nor Elrod's committee explained how the board decision violated academic freedom.

Elrod took solace in doing justice as the chair of the committee, but he used his editorial pen to criticize the board severely for inventing the leave-with-pay stratagem as a way to evade the Legion allegations. As he said, the board never stated the charges or evaluated them, resolved none of the issues, and offered no rationale for its decision. Despite Elrod's eloquent statement, the only charges of any kind focused on the quality of Fisher's teaching, not his personal opinions. Nonetheless, the committee report and Elrod's coverage in the press ignored that fact and stressed Fisher's relationship with the community newspaper. "The very important question as to what treatment faculty members shall be given on account of personal views is left unsettled."[66] In fact, the chancellor, strongly supported by the president and dean, refused even to consider the Legion charges, stating cogently that the existence of the university depended upon protecting the academic freedom of the faculty.[67] They did, however, hold that Fisher's external activities interfered with his university obligations, specifically his teaching.

Elrod argued that the decision left the faculty at risk for external engagements deemed unacceptable by the board or administrators. In Elrod's view, Fisher's partial ownership of and work with the paper remained his own business, but he did concede that faculty members had to exercise good judgment in accepting such outside engagements. In addition, he agreed that university administrators had the authority to determine if a faculty member's external engagements interfered with university duties and responsibilities, but he refused to recognize that the administrators had made that determination in Fisher's case. None-theless, he retreated immediately when critics accused him of bringing the MSTA into disrepute by manipulating the *Inter-Mountain Educator* in support of Professor Fisher.[68] Despite his protestations that he merely reported the facts of the case, within two years he had to suspend the *Educator* when the MSTA morphed into the Montana Education Asso-ciation (MEA) and created its own official organ.[69]

In both the Levine and Fisher cases, the board protected the interests of the university but respected the contract rights of the faculty mem-bers. Elrod's strong arguments in favor of academic freedom in both the reports played a role in those outcomes. In the aftermath, the board scheduled a future discussion about the outside activities of faculty members but tabled the matter before any discussion occurred during the next meeting.[70] President Clapp denied Elrod's later request for sup-port to enable the committee to survey the policies of other institutions on that issue, allegedly for financial reasons.[71] Subsequently, President Clapp used arguments similar to those Elrod had presented in the Fisher report to defend a faculty member against outside pressure for allowing unacceptable language in a student publication.[72] University policy, as interpreted by President Clapp and acknowledged by Elrod—with his caveat that faculty members had to use common sense and good judg-ment—continued unchanged into the modern period.[73]

IV

Conditions improved significantly for the university in 1920 when vot-ers approved the initiatives to dedicate one and a half mills for opera-tions and a five million dollar bond issue for construction.[74] Planning for the use of the new funds proceeded apace, but with little faculty involve-ment. To rectify this, the State University faculty endorsed a "Memo-rial" statement prepared by H. G. Merriam advocating organizational, academic, and curricular reform, specifically calling for more inclusive campus governance.[75]

In response to Merriam's questions about governance, allocation of funds, and program development, President Sisson, on 12 April 1921, appointed an ad hoc committee consisting of Elrod, C. W. Leaphart, J. H. Underwood, and J. P. Rowe to develop a structure and charge for a faculty committee on university "policy and the distribution of the budget."[76] The faculty, president, and chancellor approved the Elrod committee proposal for a standing faculty committee to make recommendations on policy, budget, planning, and any other matters referred by the chancellor, president, or the faculty. Elrod served as chair in 1922–1923 and from 1926 to 1933, with most discussions focused on the university's worsening financial situation as the national economic crisis deepened after 1930. Merriam argued that the committee kept the channels open and played a critical role in maintaining amicable relations on the campus.[77] In time, however, the faculty concept of the committee as advisory to the president changed subtly to include binding approval and veto authority, a shift never accepted by the administration.

In 1922, President Clapp articulated the administrative view when he informed the board that the committee had successfully given a voice to the faculty in university administration.[78] More directly, however, in 1926 he praised the committee as "the most effective piece of machinery that we have ever set up," but for reasons different from those cited by the faculty.[79] While the faculty stressed shared governance, Clapp viewed the committee as a mechanism to keep the faculty informed of administrative perspectives. Moreover, as he noted specifically, the committee rarely if ever offered new ideas on governance or policy. Clearly, Clapp's approach to university governance actually reduced faculty involvement in administrative work, as he assigned virtually all of the routine work to one-man committees, the university registrar, or other administrators. In doing so, he won support by releasing the faculty from administrative chores, which left more time for teaching and research.[80] He reported that those changes and the existence of the committee on budget and university policy brought a welcome end to the unproductive general faculty meetings of the past dominated by uninformed discussion of irrelevant matters.[81]

Clapp also found it worthwhile and timely that the new approach allowed him to dismiss incompetent or insubordinate department chairs and redistribute university funds without stirring faculty dissent. Overall, he concluded, the faculty members' tendency toward querulousness and conflict dissipated, and they worked collaboratively with the administration on important issues. Clapp successfully managed faculty or shared governance by consulting frequently with the committee on budget and policy—as commonly known—although consultation invariably resulted in committee endorsement of his preferences, presented as recommendations to and subsequently approved by the general faculty. His skill in coordinating shared governance combined with ever-worsening economic conditions to explain the willingness of the faculty in 1933 to approve the committee's unanimous recommendation to authorize him to suspend any rules he thought necessary because of economic conditions.[82] While seemingly at odds with the goals of Merriam, Elrod, and the faculty in 1922, that outcome manifested Clapp's administrative acumen and integrity and belied any allegation of radicalism on the State University campus.

The Merriam "Memorial" also sparked discussions about curricular reform that persisted through the 1930s and into the 1940s.[83] President Clapp used his position as chair of the university curriculum committee to lead discussions about the coherence and integrity of the education students received at the State University.[84] Clapp spoke for traditionalists such as Elrod when he deprecated excessive duplication of elementary courses, urged a coherent general education program, recommended a much-reduced number of electives carefully designed to improve student success within their chosen fields, emphasized course content over credits, and denied academic credit for practicum and internship courses impossible to evaluate academically.

With deliberate care and intent, Clapp collected and distributed reports of academic reform elsewhere and used them to propose a sweeping organizational evaluation followed by substantive change intended to improve the quality of the education and preserve the financial viability of the State University.[85] Elrod and the curriculum committee, all full professors on campus, took heart from these proposals. During the next twelve years, Clapp led a reform effort that began with incremental change and culminated in the proposed establishment in 1934 of a new organizational structure and curriculum. Briefly, he envisioned higher admission standards for entering students—designating the lower and upper academic divisions as junior and senior colleges—and a certain level of academic achievement for either transition from the junior to the senior college or dismissal from the university. Finally, students had to demonstrate a specified level of competence in order to graduate.[86] Clapp's proposals responded to national calls for reform that demanded a more integrated, structured undergraduate curriculum and the necessity for efficiency in the face of worsening economic conditions.[87]

Under Clapp's proposals, students and their advisers had full responsibility for their programs of study, with administrative work restricted to record-keeping. Threatening almost cataclysmic change if fully implemented, these proposals wrenched the State University from its traditional moorings and cast it into the troubled waters of reform and reinvention. However, Clapp died in office before bringing the reforms to fruition, and without effective leadership the effort lost its impetus. After drifting through the late 1930s and the war years, the university finally developed a modern organizational structure and a new institutional and curricular focus in the post-World War II years under the leadership of Presidents Edward O. Melby, James A. McCain, and Carl McFarland.[88] However, Elrod missed that outcome because the paralytic stroke he suffered in 1934 left him unable to speak coherently and unable to walk without the use of a cane or a wheelchair. As perhaps his last contribution, he served as the chair of the committee that designed the integrated general biology course for freshmen, a change he had advocated for years.[89]

CHAPTER FOUR

Reading the Book of Nature

Years later, Elrod wrote of his initial resistance to invitations to travel north into glacier country with its pristine lakes—Kintla, Kootenai, St. Mary, McDonald, and others.[1] He reminisced fondly of the summers he had spent doing scientific work at the Biological Station at Flathead Lake. One of his colleagues at the station, P. M. Silloway, principal and teacher from Lewistown, went north every summer after the session and annually pressed Elrod to come along. But the work around the lake and in the Swan and Mission Mountains commanded his full attention, and he declined. At the time, the Flathead Lake area remained almost pristine in its natural beauty, with very few people along the 150 miles of shoreline that he explored. Two or three times a year, he visited a special place on the Flathead River a mile below the lake for an awesome vision, "as grand and beautiful as anything I had ever seen." In 1901, he made an abortive attempt to explore Terry Lake (subsequently Lake McDonald) in glacier country, but had to turn back because of time constraints and the lack of a usable wagon road from Columbia Falls to Belton.[2] After exhausting the interesting places to study around the lake, he finally made his first exploration trip north in 1906. No longer reluctant after that experience, he returned in 1909, 1910, 1911, 1914, and many more times over the years.[3] On the early excursions, he and his colleagues collected biological specimens and traversed an area virtually untrodden except by Indians.[4]

On the 1906 trip, Elrod went to Sperry Glacier, back to Belton, to Nyack, up Coal Creek, and to the foot of Saint Nicholas, which he futilely attempted to climb.[5] Another excursion in 1909 took him to Gunsight

Pass, Lake Louise, Jackson Mountain, Blackfoot Glacier—the largest glacier on the continent outside of Alaska—and Harrison Glacier, where he spent two days in September.[6] He described the superb vista from the summit of the mountain and at Blackfoot Glacier discovered a graveyard for aerial insects. As they approached the glaciers, the roar of falling ice struck their ears "like that of a railroad train in a canyon. It was weird and impressive." The wonders of the area at times overwhelmed him.[7]

By 1909, several prominent people had succumbed to the splendor of the forest preserve designated for protection in the 1890s, despite local and national aversion to cordoning off land that might have any potential for private use.[8] But as the prospectors tired of the search for traces of nonexistent precious metals, and when the proponents of a national park compromised about mixed or joint use, the minimal resistance dissipated. Solidly among the proponents, Elrod publicly advocated a new national park.[9] A. L. Stone, the editor of the *Missoulian* and Elrod's close friend, praised Elrod's stunning photographs of the proposed park but thought the "pen picture . . . the writer has drawn" even more impressive. He considered Elrod "the man who probably knows more about the new government reservation than anybody else . . . among those who were early advocates." From his first trip into the haunting wilderness, Elrod had argued and worked for designation as the only way to preserve its magnificence for future generations.

Federal designation probably seemed straightforward to Elrod. He had relied on representative and then senator Joseph M. Dixon to secure land for the university weather observatory, the National Bison Range, and the Biological Station at Flathead Lake. Why not a national park, since the federal government already owned all of the land involved? Elrod's friends and colleagues from the bison range and other conservation efforts agreed, especially George Bird Grinnell, the father of American conservation and preservation, and William T. Hornaday.[10] Elrod credited Grinnell with the "big idea" for a national park, applauding his many years of advocacy. In fact, however, Lieutenant John Van Arsdale, United States Cavalry, made in 1883 the first public proposal to change what had become a timber preserve into a national park, and Grinnell added his endorsement in 1891.[11]

In a fragment written for inclusion in his proposed history of the park, Elrod offered a new detail to the legislative process that established the park. Elrod memorialized the claim for the record and anticipated verification in time. Contrary to all other accounts, he claimed that Senator Dixon introduced the first park bill, had it assigned to the Committee on Public Lands, and then subsequently agreed to withdraw his bill and substitute one offered by Senator Thomas H. Carter. However, Elrod provided no date for or copy of Dixon's bill. Nonetheless, his relationship with Senator Dixon afforded access to confidential information, although nothing in Dixon's papers supports Elrod's report of a deal.[12]

The legislative record indicates that Senator Thomas H. Carter introduced his bill in December 1907 and the Senate assigned it to Dixon's committee.[13] Recognizing problems with the first bill, Carter introduced a substitute bill on 24 February 1908, also referred to Dixon's committee. In late April 1908, Dixon presented a committee report amending Carter's second bill and urged its approval by the Senate. The Senate complied and sent the bill to the House. The House referred the bill to its Public Lands Committee which, at the urging of Representative Charles N. Pray, also recommended approval. Having accepted Carter's second bill, Dixon's committee and the Senate postponed the first Carter bill indefinitely in May 1908. However, the House failed to act on the second Carter bill during the session because of the pressure of business.[14] Early in the 1909–1910 session, Carter introduced a third bill, which Dixon saw through final adoption in early May 1910.[15]

If, indeed, Dixon and Carter reached agreement about sponsorship of the park legislation as Elrod claimed, they probably did so either before Carter introduced his first bill in December 1907, in February 1908 when Carter introduced his second bill, or in mid-1909, after the first effort failed, when Carter introduced his third bill. According to Elrod's explanation, Carter needed a local issue for electioneering purposes because of the developing "insurgency" within Republican ranks as Roosevelt moved toward a break with Taft in 1911–1912. Several accounts corroborate the Roosevelt-Taft break, and some Carter's struggle for reelection, but none Elrod's claim about sponsorship of the park legislation.[16] Carter undoubtedly deferred management of the legislation in various drafts to Dixon, and Dixon worked diligently with Carter and Representative Pray to shepherd the much-amended legislation through passage on the third attempt in May 1910, after successfully arranging compromises in a conference committee to settle differences between the House and the

Senate.[17] William T. Hornaday offered to launch a letter-writing campaign when the House failed to act in 1909, but Senator Dixon assured him of the bill's passage during the next session of Congress.

At the end of the day, the bill encountered very little resistance. Grinnell lobbied his many colleagues and acquaintances and reported no opposition to the bill on several occasions. Ultimately, Grinnell himself gave major credit for success to the Great Northern Railway (for the tourism potential) and the Boone and Crockett Club (Theodore Roosevelt's club for conservation and hunting). But Dixon's able management and his attention to personal relationships made the process less difficult. In any event, the enabling act specifically created "a public park or pleasure ground for the benefit and enjoyment of the people."[18] As the wording indicated, Glacier National Park manifested conservation and public use interests rather than a preservation philosophy, whatever Elrod thought.[19]

Elrod passionately urged passage of the bill in a paean to mountains, "the scroll on which is written the great story of the past," with the majesty that inspired great thoughts and noble deeds.[20] He applauded the amended boundaries that made it easier to police the park and prevent intruders or poachers from doing damage to flora or fauna. To familiarize readers with some of the park's major features, he provided a detailed description of the area based on his excursions in 1906 and 1909, noting specifically the lack of wagon roads except the rough tracks from Belton to Lake McDonald, Browning to St. Mary Lake, and the Blackfeet Agency to Altyn. During the 1909 trip, Elrod had taken the first photographs of Blackfoot and Harrison Glaciers and visited Swiftcurrent Pass and Triple Divide Peak. The area had no minerals, as the prospectors had discovered, and rumors of oil around Kintla Lake proved erroneous. While the western side hosted fairly heavy timber, he thought not much of it commercially viable, 10 percent at most, and he urged that "the Park should be set aside and maintained in its current natural condition, leaving for the future, when there is greater need, the spoliation of the timber through commercial greed," if, indeed, ever.

Very few people had experienced the park "as a playground," but he predicted thousands of visitors and supporters once people learned of its splendor. After all, he asked, "Why should Americans visit Europe for play when they have a playground at home . . . as entrancingly beautiful and inviting as [much] skill, daring, and courage as any in Europe?" Americans had only to develop their own resources to avoid wasting millions in foreign lands.[21] From every possible perspective, it made sense to present to the American people "the park in its natural state untarnished by commercial greed, and not spoiled by soulless corporations under the guise of public good." As Elrod rhapsodized, "For him who seeks rest, for him who loves nature, for him who is weary of urban life and its monotony, for him who can read sermons in stones, Glacier Park speaks God's own message."[22]

Elrod's trip to Glacier National Park in 1910 occurred just after Congress passed and President William Howard Taft signed the Carter-Dixon bill. This excursion required eleven pack and saddle horses, took four weeks to complete, and focused on entomology and the study of forests, fungi, and the lakes in the park.[23] The party traversed a circle that began at Lake McDonald and continued to Avalanche Lake and Granite Park. From there they walked the Continental Divide for miles, finding it impossible to cross Swiftcurrent Pass because of debris. They proceeded on to the Garden Wall, Grinnell Mountain, Swiftcurrent Mountain and Glacier, and Flattop, then to Waterton Lake, over Brown's Pass, past Bowman Lake, down the North Fork of the Flathead River, and back to Lake McDonald. With six people, counting himself and the guide, Elrod later reported that he and Walter Lehman walked 175 miles, as Lehman's nephew developed a bad foot and took Elrod's horse. The two of them also got separated from the party and spent a lonely and damp evening by themselves. The reunited party nearly stumbled into a forest fire near Bill Adair's place on the North Fork of the Flathead River, but managed a lucky escape after fighting fire for a time. Over the years, natural fires annually consumed an average of 30,000 acres of park timber, which, while destructive, encouraged new tree growth.[24]

I

For a proposed history of the park, Elrod wrote several chapters concerning its development, including notes on the "First Map of Glacier National Park" that identified places with and without names, listing far more of the latter.[25] In another early fragment, he discussed place names and modes of travel and predicted years before people learned about the park's special features because of the difficulty of travel. He also urged a coherent and comprehensive naming process, concerned about no place or feature named to honor R. H. Chapman, the topographer for the U.S. Geological Survey who made the first map and served as interim supervisor of the park.[26] As for travel in the park, "Pack trains have little attraction for the average traveler, are too slow and laborious, and do not give sufficient comfort to those who know nothing of such mode of travel." For the park to develop and serve its public purpose, he realized that it had to have place names, roads, and hotels, but always within strict limits in order to preserve its integrity and beauty.[27] He expected travel by foot or horseback to delay but unfortunately not prevent road dust and blaring horns.[28]

Over the years, he chronicled the construction of roads and other modes of transport, as, for example, in the *Missoulian* article about "Glacier Park's Transmountain Road . . . a Marvel."[29] While he valued the roads as avenues for people to visit the park, he preferred to keep as much of the park in its natural state as possible. Along with his friends and successors in the conservation movement, he never resolved his ambivalence concerning preservation or conservation. In an undated piece, probably written in 1932–1933, he mused about the various means of travel to the park—from horseback and on foot in 1906 to the first flight over the park of Lieutenant Nick Mamer's airplane, the *West Wind*, in the 1930s.[30] He distinctly preferred the early modes. Stated frankly as usual, he insisted upon the exclusion of all distractions from the natural wonders of the park, especially jazz, which he found abhorrent.[31] His disdain for vulgarity and commercialism only reinforced his disgust with popularly defined progress.

Despite his concern about encroachments on nature, he proposed to stock the streams and lakes with fish, to smooth the limited number of trails into roads, and to affix appropriate names to the special places for the edification of the public. In an article accepted in 1912 for publication by the Department of the Interior, he proposed names for some special places—Dixon Mountain, Dixon Glacier, Dixon Lake, Peary Lake, Nansen Lake, Haunted Lake, and Lake Louise, among others, but none for Chapman.[32] The omission did not matter, since Chief Clerk C. Reeker deleted the names Elrod proposed and substituted topographical descriptions. He explained that the board on geographical names prohibited naming park features to honor living persons and had not approved any of Elrod's names. If the board approved them prior to publication of the article, he planned to reinsert them.

Elrod demanded immediate reinsertion of his proposed names in a letter to the secretary of the interior and declined to request permission for his naming decisions.[33] Local usage had long since established most of them, and he had included a few others to honor some deserving people, particularly Senator Joseph M. Dixon, the original sponsor of the authorizing legislation. In addition, he knew of several park features named for individuals still alive, and he supported popular usage. Most importantly, "It does seem too bad to have to kill a man or wait until he dies before he can be honored." As it turned out, he prevailed on some of the names—as in the case of Dixon Glacier—but not all. However, he continued to advocate historical and popular names.[34]

To explain the exploration and development of the park, Elrod familiarized himself with the accounts of early visitors and solicited information from several old-timers.[35] William T. Hamilton, a government scout during the Indian wars in the Washington Territory, came to the Idaho Territory—including what became the Montana Territory—for three months in 1858 to assess Indian attitudes.[36] Hamilton passed through Hellgate Canyon, followed the Blackfoot River and crossed Cadotte Pass to the Blackfeet Agency on the plains, camped at St. Mary Lake, and returned over Red Eagle Pass and down Nyack Creek. Elrod also traced Raphael W. Pumpelly's trip through the area in the 1880s by reference to sites known in the 1920s, and corrected Pumpelly's explanation of glacial formation.[37] Pumpelly reported seeing the glaciers but did not inspect them. Based on Pumpelly's observations, Elrod later hypothesized that the park glaciers attained their maximum size between 1860 and 1875 and declined gradually after that time.

From Duncan McDonald, son of a Scottish trader for the Hudson Bay Company and his Iroquois–Nez Perce wife, Elrod learned of the naming of Lake McDonald for McDonald, who carved his name on a tree in a campsite at the foot of the lake in 1879.[38] (The lake, formerly known as Blaine Lake, had been futilely named Terry Lake by the Great Northern Railway for a brief time.) McDonald also visited Waterton Lake, Chief Mountain, and Kootenai Lake. He climbed Mount Campbell and later traversed Marias Pass on a prospecting trip. As an agent for the Great

Northern, Lyman B. Sperry from Oberlin College went through the park in 1895, visited the glacier that bears his name, marked the trails from "Snyder's place" (subsequently Lewis's Glacier Hotel) to various glaciers, and made other excursions into Lake McDonald country.[39]

Elrod accorded primacy to George Bird Grinnell's extensive travels through the park area after the devastating winter of 1883–1884, "the starvation period for the Indians," with buffalo bones strewn everywhere from the slaughter that left no game for the Indians.[40] As an additional project, he planned a biography of Grinnell but never expanded the speech he devoted to Grinnell's life and exploits. He also included James Willard Schultz, Grinnell's frequent companion whom the Indians named Apikuni, who first visited Blackfoot Glacier, Bullhead, Sherbourne, and Josephine Lakes, and Swiftcurrent Falls in 1884–1885, and gave Indian names to many of the park mountains, lakes, and passes.[41] Finally, Elrod added Walter McClintock to the list, a man who lived for years with the Blackfeet, developed a marvelous collection of photographs, and produced a materia medica describing the tribe's use of plants for healing purposes.[42]

Before the completion of the Great Northern Railway in 1894, a few difficult trails facilitated travel through the forest preserve that became Glacier National Park.[43] Except for emergencies, only the Indians crossed the Continental Divide for hunting, warring, or spirit quests, and they most frequently used Cut Bank, Dawson, or Two Medicine Passes following Nyack Creek to the Flathead River. Although the Indians knew the difficult trail over Marias Pass, they typically used other routes and rarely stayed long in the mountains; hunters intruded occasionally in search of the abundant game; and a few trappers, prospectors, and naturalists on scientific excursions came and went. The Great Northern Railway succeeded in increasing the traffic by improving the trails and constructing facilities. A very bad road linked Belton to Lake McDonald and a few rough paths led to other worthy destinations. "An inexpensive frame house stood where now stands the rather imposing Lake McDonald Hotel, called Glacier Hotel for years by its builder, John E. Lewis. Tourists stopped with Milo Apgar at the foot of the lake, or with Frank Geduhn at the head of the lake, in either case under the shade of the beautiful cedar trees of sweet smelling odor." Elrod's growing love for the place pervaded his writing.

In Elrod's explanation, the U.S. government purchased the eastern half of the park to the Continental Divide in 1895 from the Blackfeet Indians, whose reservation dated from the Stevens Treaty in 1855, in order to prevent hostilities between the Indians and the voracious prospectors. Thereafter, the government established the forest preserve.[44] As Elrod said, the black shale of the Cracker Mine memorialized the "abandoned hopes for wealth of the prospectors who dug the two tunnels and built the big concentrator yet standing, at great expense of money and labor."[45] While prospectors remained for a few years, and invested a great deal of energy and money in the effort, they ultimately conceded failure.

> There remains at this time, so far as this writer knows, but one miner's tent, about which grazes a lone horse, on a shelf high up on the side of Stark Point of Grinnell mountain. The Crackerjack is abandoned, the road to the mill . . . gone to ruin, the mill . . . but a source of wonderment and questions to tourists, and the old town of Altyn . . . long since . . . dragged away, log by log, until but a single cabin remains. Two holes at Iceberg lake tell of abandoned hopes. The old Bullhead mine at Mt. Wilbur is fallen in. The switchbacks leading over the cliffs at Appekunny [sic] Falls to the mines are fallen in and almost impassible. There was and is no mineral.

Hunters and trappers followed the prospectors, usually on a temporary basis. The Great Northern Railway vigorously advertised to attract passengers, in competition with the Northern Pacific Railroad and Yellowstone Park.[46] Ultimately, George Bird Grinnell, the Boone and Crockett Club, the Great Northern Railway, and conservationists across the country, including Elrod, successfully promoted a national park rather than a timber preserve. In May 1910, the area consisting of 918,681 acres (1,534 square miles, roughly one percent of the state's area) became a national park containing hundreds of lakes, mountain peaks, streams, and glaciers, with exquisite flora and fauna. Elrod vividly described the glaciers "as remnants of the ice masses of much greater size" melting annually more than they increased. He sorrowfully predicted their ultimate disappearance without an unexpected decrease in the average temperature. Nonetheless, he anticipated "moving ice in the park . . . centuries hence." To protect these cherished resources, he considered it imperative to prohibit "all utilitarian and commercial enterprises" to "save the Park for park purposes." To that end, he publicly opposed the proposed dam at Lower St. Mary Lake in the mid-1920s, using his own photographs to illuminate the destructiveness of such an atrocity in the park.[47]

II

Despite the alluring demands of each, Elrod found it possible to integrate his passion for the park with the development of the station in its new location at Yellow Bay. He and President Craig finally selected 160 acres in different locations for the station, including just over eighty-seven acres at Yellow Bay, thirty-seven acres on Bull Island, and thirty-six acres on Wild Horse Island. Elrod initially thought Bull Island offered the best headquarters site, allowing privacy for study and reflection as the lake held the encroaching farmers and fun seekers at bay.[48] The station property changed on three subsequent occasions, not counting a number of easements for power and telephone lines, fencing, and highway widening.

In 1915, A. A. White of St. Paul, Minnesota, a birder who spent time at the station, purchased three islands in the lake for $375 and donated them to the university for use as a bird reserve.[49] The three islands typically sheltered twenty-one species of birds, fifty species of trees, shrubs, and grasses, and a variety of small animals. In 1941, without congressional approval but with university acquiescence based on commitments to protect the remaining property and to revert the land if needed by the station, the state reserved about one and a half acres as a scientific and recreational park.[50] Actually, Elrod had set that area aside earlier as a public campground in a futile attempt to persuade people seeking to fish, swim, and camp to stay off the station grounds.[51] Finally, in 1944, the university traded the Wild Horse Island land for thirty-nine acres, much more useable, on the southern shore of Flathead Lake about two miles east of Polson.[52]

As early as 1904, Elrod had discussed the need for a new Biological Station. He estimated costs of at least $5,000 to construct a suitable laboratory building, $1,000 annually for two years for routine maintenance and purchase of equipment, and $1,000 for work on the grounds.[53] In 1906, he necessarily focused on collecting specimens until the new home became available at Yellow Bay, his final choice for the headquarters of the station.[54] He announced that, after completion of the construction and making other arrangements, he intended to keep the station open all summer. No longer an experiment, the station had proven itself. To that end, he simultaneously requested funds for construction and equipment from the state and from former senator Clark. To Clark, he proposed to dedicate the Clark Fresh Water Laboratory at Yellow Bay for $10,000 with a suitable dock for another $1,000. In addition, he requested funds for a large launch, equipment, and startup supplies totaling $15,000. His friends willingly sent supporting letters to Clark.[55]

Despite his focus on Yellow Bay, he remained flexible about the headquarters site in his quest for local financial support. In 1910, he wrote Emma from Polson on his way to the station, still on the Swan River with much of the equipment in storage, reporting efforts to solicit funds.[56] His conveyance by automobile as well as his description of Polson bespoke significant change during the decade since he had established the Biological Station. "Polson, the queen city of the lake, is a town of 1200 people, with electric lights, city water, street sprinkler, so they say, although not in evidence today." The town now boasted "a bridge almost completed, a mayor and other . . . officers, a policeman with a gorgeous star but no uniform and other marks of civilization, but no drunk people, no booze." Correcting himself, he had witnessed an Indian family squabble over alcohol when he strolled near the bridge. Expressing his outrage, he thought, "Hell . . . [is] not hot enough for people who would sell liquor to these poor children." Directly to the point of his visit, he told Emma of a boat trip the next day to Bull Island with some Polson dignitaries, hoping to persuade them to construct a facility for the station. His optimism proved misplaced as usual, since nothing came of the guided tour.

He also solicited supportive letters to the university president, Clyde Duniway, urging state funding for the station. Marcus Jones recommended Bull Island, not Yellow Bay, for the main site because of its isolation and easy access for water transportation.[57] Jones fairly bubbled about the possibilities with a budget of $25,000, just as Elrod had requested from Clark. Elrod's own final proposal to the president in 1910 envisioned a sound financial footing for the station instead of the hand-to-mouth existence of the past.[58] Yellow Bay had sufficient timber to construct a suitable building, while Bull Island had no useable timber, undoubtedly explaining Elrod's final selection of the headquarters site.

The Bigfork residents still wanted the station and had even launched a campaign to raise funds to support it, but he looked askance at the effort because of the scarcity and excessive price of land. The Polson people also invited the station to Bull Island, but they had no funds. At Yellow Bay, Elrod envisioned a laboratory building, a couple of boats, and ground repair and maintenance for a cost totaling $25,000, his earlier but unsuccessful request to Clark.[59] The legislature finally appropriated not $25,000 but $5,000 in 1911, but the board of examiners reduced the amount to $2,000, claiming inadequate state revenue.[60] When the State University's local executive board refused to initiate construction, the board of examiners finally released the entire amount for construction in 1912.

Even if inadequate, the appropriation allowed a new beginning for the Biological Station, and Elrod immediately planned and supervised construction of the new facility at Yellow Bay.[61] The brick building measured twenty-nine feet by forty feet and featured two stories for laboratories and offices, a concrete floor on the first level with a tile drain, a fireplace, fine woodwork, and solid hardware and locks, completed by 8 September 1912. As a result, the unfinished station hosted no scientific work during that session. Elrod acquired the necessary sinks and pipes but delayed installation of the plumbing until spring 1913 to avoid any winter damage. The plans for a boat house of thirty-six feet by twelve feet remained for completion along with outhouses and connecting trails. In his report for 1912, he celebrated the new building as a base for scientific work, with two gasoline boats for use on the lake (the old renovated *Missoula* and a new thirty-foot launch with a twelve-horsepower motor). Elrod also listed other accouterments to benefit scientific work: good equipment including tents, field apparatus, and a light canvas boat for remote lakes; a developing library; and excellent access to fieldwork.[62]

The new station offered wonderful opportunities for the study of Montana's flora and fauna, with the lakeshore to the north dotted by orchards and gardens but still relatively unsettled to the south and surrounded by beckoning mountains. In fact, Elrod thought the location ideal, about sixteen miles north of Polson over a much-improved road (one and a half hours by boat), and some twenty miles south of Somers and the Great Northern Railway. In a subsequent report, he proposed the Biological Station as headquarters for anyone engaged in the study of the state's natural history and resources.[63] However, he still needed $5,000 for salaries, a lounge, a boat dock, a small launch, a rowboat, and incidental equipment. With wonderful facilities and adequate support, he had full confidence of making Flathead Lake more productive of fish.[64]

III

Between 1913 and 1921, Elrod labored at the Biological Station but never received the support he anticipated after the appropriation for the new start. In 1914, he built a small dam on a nearby creek and installed 750 feet of pipe to provide gravity-fed running water to the laboratory and kitchen.[65] In addition, he renovated a log cabin for use as a dining hall, moved the kitchen closer to it, constructed a log boathouse, and built rough furniture for all the buildings. Clearing away the brush had put the station in fine condition, but it still needed an adequate dock with a connecting cement walk and a protective fence. Scientific work flourished, E. L. Bray on microscopic plants and protozoa, A. A. Saunders on birds, and A. G. Vestal on grasshoppers. After the station closed for the 1914 session, Elrod, once again using Clark funds, took Saunders with him on an excursion to Glacier National Park to conduct soundings on several lakes he had not yet visited. Saunders identified sixty-four species of birds while Elrod collected several thousand insect specimens, including hundreds of butterflies.

He undertook a detailed analysis of Flathead Lake based on his studies to date, projecting it to become a popular resort for tourists and a profitable venue for commercial fishing.[66] The lake had originated in a geological fault, with much higher water levels until about two hundred years earlier when the water began to fall. It had originally drained to the west until the flood cut through the Polson moraine and created the outlet to the south.[67] Thus far, he had identified nine indigenous fish species, with Columbia River chub most numerous.[68] The earlier effort to plant three million young Lake Superior whitefish had produced few if any results, since no one reported catching even one. This fish and its questionable presence in the lake subsequently caused trouble for Elrod. He expected other plantings in connected waters to increase the species available over time. To acting president Scheuch, he reiterated the request for $5,000 annually to complete the biological survey of the lake in preparation for planting new species and to conduct some experimental work with fruits and vegetables on station land. With two or three researchers each year, he predicted the ability to publish important scientific studies.[69] As usual, this request served merely to inform the president of work in progress and plans for the future, while also raising expectations about published results of the ongoing work.

In 1916, Elrod asked the new chancellor, Edward Elliott, to allow the station to retain and expend the fee revenue from students studying at the station during the summer.[70] He also proposed a refund (less $5) for each student who paid for a railroad ticket to come to the station, as granted to students who attended summer school on campus. He complained that the $375 annual appropriation failed to cover the costs of instruction and research. He also requested funding for a camp worker to handle routine tasks and release him for scientific research. The chancellor, though not a camp worker, approved some of the requests, which provided a modicum of relief. In an earlier letter, Elrod had assured the chancellor that additional support had the potential to strengthen and deepen the academic work and research, and suggested one full-time scientist in residence for the full year.[71] He also proposed eight- or ten-week sessions to accommodate the number of students and visitors expected to grow with the completion of the east shore road and the rumored construction of an east shore railway.

Although the chancellor did not acquiesce, the State Fish and Game Commission finally agreed to pay for a man to do the drudgery of lifting the water and mud from the lake bottom, releasing Elrod to search for the phantom Lake Superior whitefish.[72] In a July 1916 letter to W. M. Bickford, chairman of the commission, Elrod reported the successful planting of 3,000 brook trout in a small stream nearby. He also discovered that the squawfish suffered from tapeworm infestation, with some of the worms exceeding thirty feet in length.[73] The second most numerous species of fish in Flathead Lake, of little use to anyone, and voracious consumers of fish food, the squawfish produced a multitude of minnows, which fed the other fish. In his studies, he had identified a few additional species of fish, including suckers, a few non-native black bass, and some native whitefish. He expected to find many more species as his study progressed.

Commissioner M. D. Baldwin also opened a dialogue with Elrod in July about the Lake Superior whitefish and some hatched eggs he and others had planted seventeen years earlier.[74] Baldwin fully expected the fish, reportedly native to St. Mary and Swiftcurrent Lakes, to do well, and he sought Elrod's assistance to locate the progeny. However, he knew of no evidence to corroborate the presence of the fish in Flathead Lake. Quite common and not carnivorous, the Lake Superior whitefish lived in deep water, except when spawning in October and November. For the best results, he advised Elrod to set nets at depths of one hundred to two hundred feet at several lake locations. Because he believed the whitefish had survived, he intended to search for them himself in November, using the steamboat *Montana*. Elrod welcomed Baldwin's interest and enthusiasm, perhaps much to his subsequent regret.

At the time, Elrod assured Baldwin that he knew all about the habits, habitat, and predators of the Lake Superior whitefish.[75] However, he speculated about the different food available in Flathead Lake and the Glacier National Park lakes compared to the Great Lakes: no mollusks to speak of and few crustaceans. He had set nets at up to 150 feet all around the lake but had sighted only one fish that resembled a Lake Superior whitefish as it escaped near Bull Island. Given the lack of sightings and the probable depth the fish maintained, he had doubts about its value as a sport fish and even its presence in the lake.

In his exchange with Baldwin, Elrod exuded much more confidence than in a request he sent to a friend in Chicago, Dr. Josiah Moore, in late August.[76] Specifically, he described the search for the progeny of the three million Lake Superior whitefish, species *Chipeiformis,* allegedly planted years earlier in Flathead Lake. He had personally trapped about forty or fifty fish but did not know whether he had caught any of the planted fish. He asked Moore to secure from a Chicago fishmonger and send several Lake Superior whitefish—on ice—for comparison purposes. This curious episode revealed Elrod's awareness that he needed assistance, but he apparently received no reply. The futile search for the whitefish initially sponsored by the Fish and Game Commission continued for a few years only to end in serious conflict.[77]

For the station's fourteenth season in 1916, Elrod listed ten courses, expected adequate enrollments, and requested more university support. To buttress his request, he argued that the students already paid more than their fair share of the costs, with each student paying the regular summer session fee of $12, another $5 for field equipment, $5 for a tent for the session, and $6 a week for meals. In anticipation of a positive response, he invited Dr. Henry B. Ward to the station for the summer. As it turned out, however, the chancellor set an absolute maximum station budget for the year, well below Elrod's request, and Ward did not visit.[78]

Despite the lack of success with his requests, Elrod persisted the following year and also requested an automobile for the first time, as the completion of the east shore road made motor travel feasible. His 1917 request included funds to complete the dock and the lounge and to provide stipends for at least two national scholars to broaden the scope of research and prepare scientific publications.[79] The emphasis on funding, scientific work, fish studies, and publications generated consequences Elrod did not foresee. Disagreements within the expanded group of people planning and funding station projects had much to do with the unanticipated outcome.

IV

Elrod's 1917 and 1918 requests to Chancellor Elliott and his report for the Fish and Game Commission study delivered late in 1916 set the stage for controversy that culminated in the closure of the Biological Station for eight years.[80] In 1916, Elrod and his assistant had taken thirty-eight whitefish among the 449 fish caught in the nets, not one of them a Lake Superior whitefish. He knew about reports of some 30,000 native whitefish in the traps at Bigfork in the fall of 1915, although he privately considered the estimates wildly exaggerated.[81] As for the Lake Superior whitefish, he speculated that Baldwin and those who assisted him in planting the fish had perhaps unknowingly confused Menominee—almost impossible to distinguish from the native Williamson whitefish—with Lake Superior whitefish eggs. If so, the mistake explained the failure to capture any of the progeny. Equally critical, his meticulous search found even the native whitefish scarce during the summer, perhaps because they migrated out into lake tributaries.

Finally, Baldwin's counsel to the contrary notwithstanding, he had netted most of the natives at fifty feet or less. He conceded, however, that both the native and Lake Superior whitefish had value as sport or food fish, but he doubted a successful planting in Flathead Lake without drastically reducing the number of carnivorous bull and lake trout to protect the young whitefish. To obtain adequate evidence to validate his speculation, he proposed to extend the work season to the full year for multiple years at an additional estimated cost of $3,000 per year. These findings, speculations, and recommendations finally elicited a response, although not the one he wanted.

Elrod admitted to the chancellor in early 1917 that his report to the Fish and Game Commission required revision prior to publication.[82] He acknowledged as well that the time had come for more and better scientific publications to justify continued university support for the Biological Station. However, he warned against setting the bar too high, fearing the consequences of exaggerated expectations. As he warned, "For twenty years I have given much of the summer to the work of the Station and to field work. During all this time I have worked without extra pay except for the past four seasons." Private contributions had funded most of the fieldwork.

Elrod himself had prepared the station bulletins on his own time, and he and other scholars had published articles in scientific journals without cost to the university. For two summers, Elrod had supported Marcus Jones in the preparation of an exhaustive monograph on the botany of western Montana, likely to cost $1,000 to publish. In addition, A. A. Saunders had finished a book on Montana birds, and another scientist had nearly completed a study of Flathead Lake protozoa, the motile unicellular animals with plant-like behavior (photosynthesis) that made excellent food for nearly all fish species.[83] Elrod himself had a manuscript about "The Flathead Lake Bird Reserve" on Bird Island ready for publication.[84] Finally, only the Woods Hole Marine Biological Laboratory in Massachusetts had a higher average annual number of students than the Biological Station at Flathead Lake.

To meet expectations for the 1917 summer season, Elrod planned only advanced academic coursework and continuation of the studies on birds, fish, and fish food. He estimated the costs at $1,500, with $1,300 dedicated to teaching, but also requested $1,200 more to pay for publication of the results of the station research.[85] Following discussion with acting president Scheuch and in response to the chancellor's comment about possibly cancelling the 1917 session, Elrod reduced the request significantly.[86] He also solicited $250 from the Fish and Game Commission to continue the fish study, pledging additional research and revision of the 1916 report for publication in the commission's annual report.[87]

In a subsequent letter to W. M. Bickford, he outlined his plan of work in detail.[88] In addition, he informed the Research Committee on Fish and Fisheries of the Ecological Society of America that he had a manuscript ready for publication concerning the work in 1916 on the fish and fish food of Flathead Lake. To add to his findings, Elrod and G. B. Claycomb planned to complete a preliminary study in 1917 of the density and distribution of entomostraca as fish food.[89] As a further indication of his state of mind, he described to his brother a very productive summer of dredging to establish the relationship between the food and the fish in Flathead Lake.[90] He obviously thought the work had progressed well.

Despite Elrod's confidence, a series of letters in June, July, and August exposed the problems. The Fish and Game Commission decided not to continue the project, and Commissioner J. L. DeHart insisted that the commission have open access to all of Elrod's prior reports and data because of the commission funding.[91] Even more troublesome, Commissioner Baldwin faulted Elrod for failure to find the Lake Superior whitefish, for denying that the Richardson or native whitefish populated the lake, and for advising against any effort to plant whitefish. Baldwin stated confidently that fat and healthy native whitefish migrated by the thousands out into the lake tributaries in the fall, proving that they flourished on the food in the lake. As further evidence, he reminded his commission colleagues of taking over six million eggs from more than 30,000 native whitefish in the Bigfork traps in the fall of 1915. Yet Elrod had reported scant evidence of any whitefish, native or planted. Baldwin moved successfully to terminate Elrod's contract and to use commission employees to do the remaining work on the project.

On the defensive, Elrod indignantly denied all of Baldwin's charges. He simply had not found any evidence in 1916 of the thousands of native whitefish that Baldwin claimed to have observed in 1915. Although he had not captured any Lake Superior whitefish, he had not offered an explanation because he had no evidence. Perhaps the fish had survived, but only further research held the answer to the question. More to the point, Elrod expressed doubt that commission employees had either the expertise or the equipment to complete the study. Most emphatically, he denied DeHart's demand for open access to his reports and data. He had done the work with no personal compensation from either the university or the commission, using the commission funds exclusively to pay for the dredging labor. As he concluded, no money meant no work; no pay, no results. His expertise and integrity at issue, he bristled.

After some time for reflection, although still distraught, Elrod sent Bickford the only copy of his unrevised 1916 report *"merely for information."*[92] The preliminary conclusions revealed plainly that he had not opposed carefully planned efforts to plant more whitefish, nor had he denied that the lake offered adequate food to support more plantings.[93] Instead, he had urged more study as the logical step toward the development of a systematic plan for future plantings. When the commission held firm, he decided not even to pursue reimbursement for completed work because of the unfavorable decision already rendered.[94] He pledged, however, to complete the revision of his report for publication, excising any extraneous matter, and promised Bickford a reliable publication complete with his findings and recommendations.

Although only a preliminary study, Elrod's initial funded project concerning the fish and fish food in the lake ended on a very sour note. He had rarely encountered hostility of this sort, perhaps not since the retirement of former president Craig in 1908, and he did not handle it well. In response to queries for information about fish-related industries in Montana, Elrod disgustedly replied that the research proceeded very slowly for want of support. "At present we have no idea of the amount of life in the waters, and trust to good fortune in planting fish from year to year."[95] At the time, further study seemed very unlikely.

As one shoe dropped, the other quickly followed, and Elrod came under severe criticism for failing to deliver on his university commitments. Chancellor Elliott and President Edward O. Sisson demanded scientific results, specifically publications. Once again, Elrod reacted petulantly and resentfully. Instead of a careful listing, he sent a review of the Biological Station's development, either to remind or to inform every one of the facts.[96] He, not the state or the university, had established and nurtured it as a labor of love, supported primarily by private contributions and the effort and expertise that he and visiting scholars donated over two decades. Moreover, the station had not just survived but actually flourished, with neither state nor university support. Only in the last four years had anyone, including himself, ever received compensation for station work. The researchers typically paid most of the costs of their research, and even the students paid fees for transportation, equipment, instruction, and room and board. The state had belatedly provided construction funds for the laboratory at Yellow Bay, but only after Senator Dixon, at Elrod's request, persuaded the U.S. Congress to grant the land free of charge.

The publications had cost the state and university nothing, since the researchers did the work without pay and Elrod handled the arrangements. To demonstrate the extent of the work, he provided a partial listing of the articles in scientific journals and university bulletins as well as some completed manuscripts.[97] Subsequently, he lackadaisically tried to develop a full list of station publications but never quite succeeded. This book's appendix contains a composite listing that reflects work done between 1900 and 1926, most of it by 1920. Elrod's feeble attempts to provide a complete list never satisfied the administrators, either because he lacked the necessary information or because he thought the demand inappropriate and insulting. In any event, he devoted less than full attention to what he regarded as an obnoxious demand, and the station suffered rather severe consequences as a result. The appendix draws on several of his perfunctory responses.

V

Despite his intransigence, in early 1918 Elrod seemed confident that he had assuaged the concerns of the two administrators. In March, he submitted a Biological Station budget request for $1,200, the same as the prior year.[98] Chancellor Elliott reluctantly approved $1,050 plus $250 in student fee revenue, citing the fiscal pressures of World War I, the need for economy, and insufficient scientific results. While he appreciated Elrod's years of unrewarded effort and understood that prior budgets hardly warranted serious scientific results, he had frequently requested finished manuscripts but Elrod had provided nothing ready to print. The chancellor reminded Elrod of his comment in 1917 that the report for the Fish and Game Commission did not merit publication.[99] In his funding approval for 1918, the chancellor warned Elrod that future support depended upon evidence of clear progress by 1 October.[100] The warning either failed to impress Elrod or his report took more time to complete than he anticipated, since he missed the deadline and suffered the consequences.

President Sisson underscored the chancellor's demand for progress as the price for future funding for the station.[101] Because of his own recent arrival on campus, he had no background information as a basis for judgment and thus no foundation to argue with the chancellor in support of Elrod. To inform himself, but without notifying Elrod, Sisson solicited counsel about the station from informed professionals. Professor of botany Paul W. Graff, who had worked at the station under Elrod's direction, responded with recommendations for increased effectiveness and efficiency.[102] He considered the Yellow Bay site superior to the one at Bigfork, having worked at both. He also thought the numbers of students and scientists about right but questioned the quality of the students. Graff urged a requirement for all students in relevant fields at the university to spend at least one season at the station enrolled in more than one course at a time. He also recommended only advanced courses and at least one in geology. Finally, and potentially most damaging to Elrod, he faulted the paucity of publications.

Unaware of Sisson's inquiries, Elrod requested $1,200 for the 1918 season, with $750 for teaching.[103] He repeated the litany about building the station and its excellent plant as a labor of love with little other than private donations. He sorrowfully observed that despite his accomplishments, neither the chancellor nor the president actually supported the station. Tardy decisions and inadequate resources year after year had convinced him of the futility of devoting time and worry to it. Far more prudent to teach on campus and avoid the inane criticism of a "vacation at the expense of the state." Deep disappointment underscored his words: "I really have lost inspiration and courage for the work at the station." Announcing his willingness to undertake any assignment likely to earn approbation, he asked only for clear direction. "I am not wedded to the station, although I see great possibilities and opportunities there."

Wearily, he closed with a review of the work on fish and fish food during the last two seasons and his plan to continue in 1918.

Elrod's report on the 1918 session of the Biological Station arrived unusually late, in mid-August 1919, delayed perhaps because of revising the Fish and Game Commission report he had completed for Bickford.[104] Perhaps he had bogged down as well in his effort to identify publications to report to the chancellor. Even if late, however, the 1918 report reflected his intent to demonstrate worthy progress. Graff had studied the microscopic plants in the lake; G. B. Claycomb analyzed how gas, light, and temperature affected the distribution of fish; and Elrod personally took soundings—and collected and bottled specimens and food samples—at multiple locations around the lake and in adjacent lakes and streams. Based on this work, he provided drawings of twenty-two species of entomostraca and fifty species of diatoms (phytoplankton), a common form of algae.[105] Finally, he also reported data on the lake's rise and fall with daily observations over two seasons, and compared evaporation rates from the lake's surface with those from within the dense woods around the lake, having placed evaporators and thermographs at two locations for twelve weeks.[106] He pledged to send the final reports on all the work as soon as possible.

Unknown to Elrod, President Sisson had received a confidential report from T. C. Frye, director of the Puget Sound station, in early 1919 with specific recommendations concerning the Biological Station at Flathead Lake.[107] As Frye observed, biological stations typically enrolled few students and did not impose heavy teaching loads on the faculty because of the research requirement. As a result, the cost per student usually exceeded that on campus, well worth it because of the research benefit that accrued to the students and the university. For his station, Frye estimated the cost of research publications averaged about $1,000 per year in a total budget of roughly $3,700, with $600 generated by his station. More specifically, he observed that the most effective stations enjoyed access to both fresh and seawater environments, and he thought the Flathead Lake station severely limited because of that disadvantage. In fact, he saw little value in the Flathead Lake station. Frye recommended saving money by closing it and subsidizing students and faculty to study and conduct research at the Puget Sound station. Without expressing his own opinion on the merits of the recommendations,

Sisson shared both the Graff and Frye reports with the chancellor. Out of concern for Elrod, however, he urged the chancellor to fund the station at least for the 1919 season.

Elrod chafed under the administrative demands. Making a bad situation worse, he received disturbing reports in 1919 of vandalism at the station. In May, a friend found a stranger from Kalispell occupying the dining room, Polson boys and girls cavorting in Elrod's laboratory, and considerable damage to the building.[108] In July, Acting Dean of Forestry Thomas Spaulding advised Elrod to take steps immediately to protect the station from fire.[109] He had found eighty people occupying it on the Fourth of July, throwing firecrackers everywhere despite extremely dry conditions with no thought of the consequences. He had stopped these antics but predicted recurrence. He specifically warned of the danger threatening the station's facilities and equipment in view of the looming fire season, likely to become the most severe since 1910.[110] Spaulding doubted the efficacy of fences or gates and recommended security guards to keep out the boatloads from Polson, Somers, and Bigfork.

Problems arising from crowds on the premises hardly surprised Elrod. He undoubtedly anticipated as well the likely response to a request for a security guard. Chancellor Elliott predictably denied any relief but called on Elrod to develop a plan for the protection of the station when unoccupied, although offering no assurance of funding it.[111] In the same letter, the chancellor lamented the paucity of published or publishable manuscripts from Elrod, claiming he had received nothing. Maybe the president had them. In all likelihood, Elrod had yet to locate and send the list of publications. Probably as a result of these undercurrents of controversy, the station opened for only one week in 1919 to accommodate a few forestry students.[112] Whether Elrod ever learned about or had access to the Graff and Frye reports remains unclear. No evidence suggests that he did. Be that as it may, tenuous relations with the two administrators, the alerts about the lack of security at the station, and the abbreviated 1919 session certainly conveyed cautionary signals. Yet he apparently gave no indication that he anticipated anything other than business as usual in 1920.

On 20 June 1920, Elrod confronted the consequences of his procrastination and intransigence when a *Missoulian* headline proclaimed, "Shy of Funds; Station Closes."[113] For the first time since the new beginning in 1912, the station closed for lack of funds, according to the chancellor. With some exaggeration, Elrod lamented that for eleven sessions after its founding in 1899 the Biological Station had attracted large numbers of students to the Swan River site. With Joseph Dixon's success in securing the grant of reservation land and a welcome legislative appropriation in 1911, the station had experienced a decade of robust development. As significant accomplishments, Elrod cited the bison range in Ravalli and the Flathead Lake Bird Reserve, praising as well the accumulation of research data on the flora and fauna of the region. Under his direction, the station had promoted scientific explorations from Glacier National Park to the Canadian boundary and beyond, and had attained stature as "the only place in the Rocky Mountains where without pay and with little encouragement, the innermost secrets of nature were sought by devotees of science." His choice of words bespoke his disgust with the decision; still he entertained no doubt that the station had enhanced the university's reputation. After surviving the war years, he deeply regretted "that the station must be closed for lack of funds." The frustration and intransigence evident in his attitude and comments had unquestionably influenced the decision to close the station.

As always, however, he put the best face possible on this unwelcome development. In retrospect, the loss of Fish and Game Commission funding, his own dilatory response to the chancellor's requests for publications, the sharp Graff critique, the negative Frye recommendations, and the security problems all combined to make the decision to close for at least a year relatively easy. Nevertheless, the public approval in 1920 of the one-and-one-half mill levy exclusively for higher education operations and a $5 million bonding bill for construction promised relief for budgets hard-pressed during the war years.[114] The question for Elrod had to do with the disposition of the new funds.

He continued to work at the station through the 1921 session, although he complained that a late decision that year reduced enrollments and visitations. Despite the delay, the station attracted thirteen students and three instructors, with seventeen people working in the field and the laboratory.[115] Station studies that year focused on birds, botany, zoology, ecology, butterflies, and fish and fish food, with five excursions into the Mission Mountains and numerous boat trips around Flathead Lake. The session lasted only six weeks despite Elrod's stated preference for nine. He had found the buildings, grounds, and equipment in excellent condition, unused since 1918, with only the need for minor repairs and maintenance, replacement of a few small items and the two relatively old boats, and more tents and bed springs. This optimistic report suggested that he anticipated receiving some of the enhanced university funding for the station.

The dark clouds continued to gather as he discovered intruders again in search of a pleasant place for summer fun. In fact, they intruded even during the session. While he always welcomed visitors interested in the work of the station, he knew that these people had no time for nature study or research. They came only to lounge, swim, and fish, and they left their litter behind when they departed. Even with a small camping area free to the public less than a half mile away, they tramped through the station looking for cleared space and a good beach. Elrod had erected signs and installed a gate with barbed wire on each side, locking it when no one occupied the station, but nothing worked to keep them out. The station had to have better security and fire protection because of the value of the university property.

To that end, he hopefully proposed a year-round live-in caretaker. Once again, he recommended extending the session to nine weeks, inviting eastern specialists to conduct investigations, and additional funding to publish the studies. He reported regretfully that Marcus Jones had withdrawn his authoritative work on the botany of western Montana because the university had failed to publish it. He seemed oblivious to the recent deterioration of both relationships with the administration and the physical conditions at the station. Instead, he as much as demanded an early decision about the 1922 session in order to assure good attendance. He commented optimistically that travel along the eastern lakeshore in a buggy light enough to lift over trees and boulders in 1912 had given way to a highway with literally hundreds of cars destined for Glacier National Park in 1921. In his opinion, the new road from Bigfork along the Swan, Clearwater, and Blackfoot Rivers to Missoula augured even more traffic, enhancing the potential of the Biological Station to attract great numbers of visitors. Undiminished, and against all odds, Elrod's optimism proved resistant to reality.

In 1921, he and Francis Ross completed their most comprehensive study of the Flathead Lake fish.[116] The 1916 study had identified nine species in the lake, but the new report listed eleven more, mostly the results of authorized or unauthorized plantings. The rainbow trout, as distinct from flat trout, planted in the lake years earlier, had become numerous by 1921. Bluegill sunfish, introduced in Flathead County in 1910 with 500 fry in Church Slough, had spread throughout the interconnected water systems, including almost every lake and pond in the region. The plantings in Church Slough included crappie, small-mouth black bass, and bullheads, all of which proliferated. More recent plantings in the lake itself introduced Chinook salmon, King salmon, and Lake Watcom salmon, all very numerous. Since the salmon typically died in five years, their presence in great numbers around the lake proved they had done well. The landlocked salmon planted in 1913 in Whitefish Lake had migrated to Flathead Lake, and Elrod predicted a huge increase in their numbers. Despite this favorable report about new and different species, no one had yet caught or netted any progeny of the three million Lake Superior whitefish planted several years earlier, Elrod's bête noire of 1917. Conceding without evidence that the fish survived, Elrod suggested that slow adaptation augured more time for the fish to establish itself. After all, no one found a trace of the black bass planted in the lake during the 1890s until 1914. In time, he expected the Lake Superior whitefish to do well. His earlier error simply allowed too little time after planting for the fish to propagate. He ended his 1921 report with a recommendation for continued funding to develop a comprehensive survey of the lake's fish population and an accurate estimate of its carrying capacity.

A handwritten fragment in the folder with the 1921 report compared the Biological Station at Flathead Lake with others around the country. Elrod proudly praised the excellent work done over the years by students and researchers, briefly discussed again the station's historic development, and identified some of the eminent scientists who had taught or investigated at the Swan River and Yellow Bay sites.[117] Perhaps he had finally seen Frye's 1919 recommendation to close the station, but he did not mention it. Even with his continued faith in the station, he voiced weary disappointment and bitter frustration that his labor of love had elicited neither support nor recognition. In response to the recent criticisms, he pleaded with friends of the station to come to its and his assistance. As he remarked plaintively, the administration denied that the results obtained at the station warranted the investment of $1,300 minus the $250 in student fees. He urged all who had worked or studied at the station to write to Chancellor Elliott about their rewarding and productive experience. Belatedly, he also called on the visiting scientists to publish their work and give credit to the station. He finally understood that *only* publications sufficed to persuade the authorities to keep the station open.

After a quarter century of rare opportunities for nature study, science education, and research at the station, Elrod concluded painfully, on the basis of his experience, that students and their education had never really mattered all that much. He at last came to grips with the university administration's attitude toward the Biological Station, and he fully recognized the implications. A former student wrote in 1921 to voice concern about the closure of the station at Yellow Bay for no good reason.[118] Robert Oslund praised his experience at the station, and he considered the training he had received comparable to that of students at Woods Hole and other stations. In the end, Elrod's analysis of why the station closed agreed with that of the chancellor. The administrative dictate made it clear that the quality of the student experience did not matter.

As always, however, the chancellor had the last word and Elrod accepted the inevitable. Rather than waste time and energy about matters beyond his control, he placed the blame elsewhere and sought new opportunities. In 1921 the administration of the university changed as well when Sisson resigned to return to teaching at Reed College in Oregon.[119] Elrod learned of Sisson's resignation while recruiting high school students around the state, and he wrote Emma of his surprise.[120] His own experience in 1908 led him to suspect a deeper reason for Sisson's resignation. Knowing as he did of Sisson's defense of faculty against charges leveled against them by the chancellor, and aware of the former president's deep interest in scholarship and teaching and distaste for administration, Elrod in all likelihood felt keenly the loss of an ally in the struggle to sustain the university and the station.

However, he never lost his bearings, and instead asked Emma's opinion in March 1922 about a new adventure. He had discussed an intriguing opportunity for collaboration with the National Park Service to conduct research and to provide interpretative services for tourists in Glacier National Park.[121] As Elrod put it, the plan called for the university professionals "to be the persons who open the doors to the visitors and make necessary explanations." The interpreter duties included delivering lectures; preparing exhibits of the park's geological features, flora, and fauna; escorting tourists on hikes and horseback trips; providing promotional and descriptive pieces for placement in the public press; and coordinating the services. The stipend of $100 a month for two or more months during summers appealed to him, and the Park Service and Glacier Park Hotel Company agreed to pay his room, board, and transportation within the park. He, two of his colleagues, and new president Charles H. Clapp had planned the project and welcomed the opportunity to work in the park with expenses paid and stipends. But first he wanted Emma's opinion, since he had secured permission for her to visit him for brief periods at no expense except travel to the park, just as she had for most years at the Biological Station at Flathead Lake.

In seizing this opportunity, Elrod opened a new chapter in his life as a naturalist-educator, but he never abandoned the Biological Station. Before the close of the first season in the park, he wrote Emma reaffirming his strong sense of ownership. A teacher had requested permission from President Clapp to house some vocational students at the station for two weeks of summer instruction.[122] Elrod strenuously objected to the request and urged the president to refuse permission. If the project moved forward, it undoubtedly meant more litter and debris and possible damage to the station. More importantly, he planned a week in September at the station for himself and Emma and clearly wanted no interference.[123]

Chancellor Elliott also resigned in 1922, to become president of Purdue University. Perhaps Elliott's departure sparked new hope for an immediate return to the station. If so, a late decision about opening, combined with the theft of some equipment, left the hope unrequited and Elrod free to return to the park.[124] Eager to develop the Park Naturalist Service he had conceived, Elrod spent eight glorious summer seasons in another labor of love.

VI

Before he began to develop the proposal for a Park Naturalist Service, Elrod made two more scientific trips to Glacier National Park. In 1911

the Elrod party went initially to Lake McDonald, on to Flattop, Chaney
Glacier, and Waterton Lake, and then spent a week at Brown's Pass
studying the geology and flora.[125] They had planned to go to the Hole-
in-the-Wall, climb to Agassiz Glacier, and ascend Mount Cleveland, but
severe weather forced a return to Flattop; from there to Granite Park,
Swiftcurrent Pass, Iceberg Lake, Grinnell Lake, Babb, Gunsight Lake and
Pass, Sperry Glacier, and finally Lake McDonald, a three-week excursion. The 1914 expedition started at Glacier Park Hotel on Lake McDonald, went to Two Medicine Lakes, Spot Mountain, Bison Mountain, Cut
Bank Creek, St. Mary Ridge, and Red Eagle Lake, back to Upper St. Mary
Lake, to Going-to-the-Sun Camp, over Piegan Pass to Many Glacier at
Swiftcurrent Lake, to Iceberg Lake, back to Upper St. Mary Lake, and
finally, by automobile for the first time, back to Glacier Park Hotel. In
his various trips, Elrod had covered most of the park, except for the Belly
River region, and he soon had that opportunity as well.

On all of his expeditions, he carried along a Kodak camera (3.25 in. by
5.5 in.) and a plate camera (6.5 in. by 8.5 in.) with four lenses, altogether
weighing about forty pounds and protected in a watertight pack. Each
excursion required twelve to fifteen dozen plates and twelve to twenty
rolls of film.[126] In a Department of Interior publication, Elrod described
his method of studying fourteen lakes during 1909–1911, using a plumb
bob and an apparatus for measuring telephone wire during installation
to gauge depths. Other essentials included a small bucket with a very
fine silk screen to capture the microscopic fish food, as well as a collaps-
ible canvas boat.[127]

Despite the scarcity of food in the lakes, he thought fish culture so
worthwhile that he stocked the lakes. His persistent optimism about the
park lakes colored his comment in 1925 about a trip to Hidden Lake to
learn what happened to the 400,000 eyed-stage eggs he had planted two
years earlier.[128] He refused to consider that the fish had not survived,
despite the lack of any evidence to the contrary. Much the same had
happened at Lake Ellen Wilson with no fish found until four years after
a planting. Because Hidden Lake had ample fish food, he anticipated
positive results. During the years he served in Glacier National Park,
he continued to dredge the lakes to ascertain the availability of microscopic organisms to provide food for minnows. As he explained, numerous high falls isolated the lakes, forced reliance on planted fish in the

absence of any native fish, and resulted in a minnow supply inadequate to sustain large fish populations. However, in time he found that the quasi-sterility and extreme fragility of park lakes inhibited fish production except in the larger and less isolated ones.

Elrod had learned a good deal about the park from his earlier explorations, research, and cooperative work with the Great Northern Railway to secure passage of the park legislation in 1907–1910. Not a conservation-minded enterprise, the Great Northern supported park designation as the means to entice more tourists to use the train for access to the wonders of the west. The Northern Pacific Railroad had pioneered the model with Yellowstone Park, which by 1915 had over four hundred miles of improved automobile roads.[129] Even earlier, the Great Northern had requested permission to use George Bird Grinnell's "Crown of the Continent" article for promotional purposes.[130]

During the following decade, the Great Northern relied heavily on a "See America First" campaign, a theme Elrod borrowed in his own articles and lectures about the park, to lure more cash-carrying tourists to the mountains. In addition, the outbreak of World War I, along with two international expositions in 1915—one in San Francisco and the other in San Diego—not only brought more people west but also contributed significantly to the demand for improved roads to accommodate automobile traffic across the country and within the park.[131] The first two transcontinental roads date from 1912, and auto tourism became increasingly popular during the second decade of the twentieth century. After the adoption of the National Park Service Act in 1916, Director Stephen T. Mather joined the Great Northern in the effort to get tourists into the national parks. Founding Glacier National Park superintendent William R. Logan, who had accompanied Lyman Sperry in 1895 through the area that became the park, enthusiastically supported the effort.[132] Nonetheless, a relatively well-developed road system did not exist in the West or in Glacier National Park until the 1920s. Elrod's plans for a park project had to await healthy tourist traffic for success.

In 1912, Elrod inquired about the Great Northern facilities, those already constructed or under construction, and the transportation and other costs to tourists. H. A. Noble, then general passenger agent for the Great Northern Railway, described the Swiss chalet–style hotels available in 1912 at Two Medicine Lake, Cut Bank Canyon, St. Mary Lake,

The Narrows, Gunsight Lake, Sperry Glacier Basin, and Lake McDermott (Swiftcurrent Lake), with more planned.[133] Noble also identified the two major entrances to the park—Belton and what became East Glacier—and the few existing roads and trails, and he provided hotel prices and rail fare from Minneapolis ($35 round trip). Elrod had already imagined the research and income possibilities in the park, and he continued to discuss them with colleagues, park, and hotel personnel until they reached an agreement in 1922.

To promote tourism, Superintendent Logan built three miles of road in 1910–1911, including a macadamized road from Belton to Lake McDonald. He also installed a telephone line to Lake McDonald and extended it to St. Mary Lake, and cleared and improved the trails to Swiftcurrent, Red Eagle to Nyack, to Cutbank, and to Two Medicine. However, the Great Northern did far more construction of facilities as well as roads and trails, with the encouragement and authorization of the National Park Service. Director Mather observed that "scenery is a hollow enjoyment if the tourist starts out after an indigestible breakfast and a fitful sleep on an impossible bed."[134] The railroad and the Park Service seized the opportunity, as did Elrod when conditions allowed.

Under Louis Hill's supervision, the Great Northern constructed Glacier Park Lodge in East Glacier in 1913 and Many Glacier Hotel in 1915, at a cost of $500,000 each, and put the hotels and chalets under the management of the Glacier Park Hotel Company, a Great Northern subsidiary. In addition, the company continued the development of the chalets about a day's horseback ride apart. W. N. Noffsinger organized the Park Saddle Horse Company in 1915 to serve the Great Northern's facilities and the tourists, by 1925 transporting 10,000 people per year with more than 1,000 horses. Roads connected the rail terminal at East Glacier with Many Glacier and beyond by 1917, as the federal government provided $534,000 and the railway $1.5 million to develop roads and trails, for a combined total of 841 trail miles alone by 1930. An improved road connected Belton and Apgar in 1917, extended by 1926 to Avalanche Lake, with plans to continue it as the Transmountain or Going-to-the-Sun Highway to Logan Pass and on to East Glacier and Many Glacier, at a cost of roughly $3 million. As roads and trails expanded, the Park Service added public campgrounds placed strategically alongside them.[135] The famous red buses operated by the Hotel Company soon captured public attention. From about 4,000 in 1911, the number of park visitors increased to 12,000 in 1913, 40,000 in 1925, and 75,000 by the end of the decade. The rapid influx of visitors opened new opportunities for entrepreneurs ready and willing to serve their wants and needs.

VII

Elrod's inquiry in 1912 suggested that he had caught a glimmer of income possibilities in the park. He understood that he had to wait for an opportune time to develop a project. John E. Lewis operated a private hotel at the head of Lake McDonald, and Frank Geduhn had a small inn there as well, until the Park Service acquired those facilities in the 1930s. M. P. Somes received permission from the Park Service to initiate a Nature Guide Service on a fee basis near Many Glacier Hotel in 1921.[136] However, the joint project Elrod initiated in 1922 with the university, the Park Service, and the Glacier Park Hotel Company provided free interpretive services for tourists throughout the park and forced Somes out of business by August of that year.

Elrod's contract to develop the Park Naturalist Service required him to spend time at Many Glacier Hotel, Going-to-the-Sun Chalet, and Lewis's Glacier Lodge, the first two managed by the hotel company.[137] He found the accommodations at Many Glacier Hotel quite comfortable. Although the company typically required the occupants of adjoining rooms to share a bath, Noble ruled otherwise and also declined Elrod's offer to pay.[138] As a result, he enjoyed a good-sized room and bath to himself, found the food decent, and had free access to a horse at any time. His base at Many Glacier Hotel placed him strategically to interact with the great flow of tourists into the park by way of the railway and the company buses. As it happened, he welcomed Emma and daughter Mary for a week visit during that first summer.

Elrod found the volume and value of park traffic astonishing.[139] The horse outfitters made about $600 per day, the company buses averaged $1,000 a day, and the hotel income challenged his imagination. He certainly had judged the potential for income correctly. That summer he made trips guiding tourists over Swiftcurrent and Gunsight Passes, and established a pleasant friendship with the artist in residence located down by the chalet, Oliver Dennett Grover from Chicago, engaged in painting the falls and Grinnell Glacier.[140] Elrod thought humanists in the park inexplicable. In addition to the painter, who subsequently went to Belly River, the poet Vachel Lindsay held forth at Going-to-the-Sun, with all expenses paid by Great Northern for a year in the Davenport Hotel. As he pondered these practices, Elrod confessed that "I don't quite see how they expect to get their money back from Artists and Poets."[141] Only scientists, professional interpreters, guides, bus drivers, horse wranglers, experienced mountain climbers, and tourists in the park made sense to him.

In August of the first year, Hotel Company manager H. A. Noble and Park Superintendent J. R. Eakin asked Elrod to accompany some tourists on a deluxe trip to Belly River and to prepare descriptions of the trails in the Belly River region, another part of his duties.[142] A month earlier, he had informed Emma that he had already written descriptions of five trails and had a story about walks around Many Glacier in progress. He complained about the slow typing and mimeograph service, but he had completed descriptions of Apikuni Trail, Iceberg Lake, Swiftcurrent Pass, the new Altyn Trail, and Grinnell Glacier. Tourists informed him they relied on his stories and descriptions to enhance their understanding of the park.[143] Park Superintendent Eakin confided Noble's satisfaction with his performance and that they wanted him to continue his service for years to come.[144]

The Belly River trip usually took several days and required wranglers, guides, tents, beds, cooks, horses, photographer, correspondent, and a naturalist.[145] A typical breakfast on the trail consisted of bacon, potatoes, cream of wheat or corn flakes, pancakes, jam, cantaloupe, oranges, coffee, tea, cocoa, and more.[146] On the fifth day out, Elrod boasted that he now knew the entire park like a book, having never before traveled to Belly River. That evening, he remained in camp with only the cook's helper, as all the others had gone to Waterton for a dance. Entranced by the trip, he seized the opportunity to describe in vivid detail an encounter they had with a sow grizzly and two cubs just after leaving Ranger Lou Sarratt's place on the Belly River:

> There were nine of us on horseback lined up in the narrow trail in the brush, Sarratt ahead, I was next. I did not know the creek was just ahead, and did not see that Sarratt was about at the bank. He called

me to hurry and see the grizzly bear. She had two cubs, and . . . I saw her as he spoke, when she heard his voice, she raised on her hind legs, looked around, located us, dropped on all fours, and with her head wagging from side to side came at us on a lope. Don't need to say they will not charge without provocation when they have cubs. We could see both cubs, little fellows, fortunately they were on the other side of her. S. yelled to turn the party and run, she was coming. I lost no time you may be sure.

My lead ran out, and it was so dark I could not put in another. I yelled at them to get, a grizzly was coming. Some obeyed with haste, others not knowing grizzly, wanted to see her, and moved slowly or not at all. Sarratt and I both expected a bear in the bunch of horses but she stopped at the creek, and then went off through the bushes with her babies. Some of the people realized the situation, but some had no idea of a grizzly except they wanted to see her. S. and I talked it over yesterday. It was all done in a few seconds, but we both agreed we did not want another trial of her, and that she meant business.[147]

Not every outing proved as exciting, but all offered dazzling scenery and opportunities to photograph glaciers, mountains, and flowers and to educate tourists. Periodically, Elrod urged Emma and Mary to come for brief visits, since the crowds thinned in late August. Meanwhile, he completed more stories, several on the park, Indians, animals, flowers, glaciers, and trails for Superintendent Eakin, sending many to the *Missoulian,* other newspapers, and the National Editorial Association.[148] While he had not seen any of the pieces in print, the superintendent assured him of their value and usefulness. His only problem derived from lack of time to write.[149] Exemplifying his diligence, he mentioned to Emma in 1928 his twenty-fifth story for that season, each about 750 words long.[150] As a direct result, his reputation for interesting stories spread as his time in the park extended, and he broadened the range and scope of his repertoire. University president Charles H. Clapp and family also visited him on occasion, and during the visits they shared enjoyable hikes to various special places.[151] In late August of the first year, he wrote of a fine summer and informed Emma that he intended to leave the park on 1 September, having earned $240. "What shall we do with it now?"[152]

During his first two years in the park, he researched and wrote *Elrod's Guide and Book of Information of Glacier National Park,* an authorized but privately funded book.[153] Hotel manager Noble declined a joint venture because of company policy but predicted success for the book.[154] In addition, he considered Elrod uniquely qualified to write a guidebook, pledging his private assistance in the venture. Consisting of 200 pages of text, maps, and photographs, the book's readability and accuracy enhanced its usefulness and attractiveness to tourists. (He expanded the 1930 edition to 258 pages.)[155] In all respects, he intended to assist tourists, hoping thereby to increase sales, and included graphic and lovingly detailed word pictures as well as photographs of trails, flora, and fauna. He counseled that "the vanishing big game animals may be seen by all with a little care," as well as the "wonderful, shining mountains . . . never twice alike" but always in sight. Elrod's fascination with and love for Glacier National Park gave character to the book.

In the book, he identified Mount Apikuni, named for J. W. Schultz, the early park explorer, using his Indian name, and visiting Chicago artist Oliver Dennett Grover who spent a summer painting at Many Glacier and thought Cracker Lake "the most beautiful trail, the most beautiful lake, and the finest combination of colors he had ever seen." Elrod also informed his readers that the black shale surrounding the abandoned Cracker mine offered silent witness to the futile search for precious metals. A photograph introduced George Bird Grinnell, taken in 1926 when he and Grinnell visited Grinnell Glacier together. He obviously took greatest pleasure in calling attention to a narrow valley along Cataract Creek near Morning Eagle Falls and the towering Garden Wall, surrounded on three sides by protecting mountains, where the trail opened into a dazzling flower garden Elrod named "The Garden of Heaven."[156] His fascination and love for the park illuminated the guide for tourists.

Elrod's Guide alerted tourists to park rules, regulations, entrances, roads, and facilities; the appropriate attire and behavior to assure pleasant hiking and camping experiences; how much food per person to carry each day; and even which fish to eat and where to catch them. Elrod included a prideful description of his own Naturalist Service, designed to meet real needs and free to the public. He boasted that he had developed the service from nothing to a high level of activity, with one assistant at Belton and others planned for various sites around the park to maintain exhibits, deliver lectures, and guide tours.[157] Widely known across the country because of its reliability, *Elrod's Guide* made him famous as the authority on the park. It quickly became the bible for thousands of tourists, and the superintendent proclaimed it "the most attractive park publication I have ever seen."[158] Slow at first, sales jumped in 1925, and Elrod anticipated selling an average of a thousand copies a year.[159] He did a sales analysis in 1930 for the first edition, with gross income of $3,193.46 at a cost to him of $2,075.66 and losses of $175.50, netting $957.84.[160] The second edition hit the stands after the onslaught of the Depression but unquestionably did well. Many of the numerous descriptions of park trails, passes, lakes, and glaciers also reappeared in various brochures issued by the Park Service, the Glacier Park Hotel Company, and Park Saddle Horse Company.[161] In 1925, he told Emma that someone, probably Superintendent Eakin, had urged him, as the most qualified author, to write a history of Glacier National Park.[162] He intended to comply but unfortunately never completed the project.

Ever alert to the main chance, Elrod took advantage of free time in the park to conduct research on the plants, particularly the flowers he arranged in wonderful exhibits for the tourists. On one occasion, he walked Piegan Trail for seven miles to collect flowers, then spent the afternoon labeling them.[163] On another, he searched through the woods around the Granite Chalet for the camping site on one of his early excursions, but failed to find it because of the changes wrought by time. He had to settle for beautiful flowers and photographs.[164] Responding to another suggestion in 1925, he launched a project for a book on the park flowers, with the only available work obsolete and cumbersome for tourists to carry around. Over the next few years, he gathered specimens and photographs for the book.[165] In July 1928, he boasted to Emma that he had completed another chapter.[166]

W. H. Mills, advertising agent for the Great Northern Railway, proposed that the Great Northern publish the book and pay royalties to Elrod of fifteen cents a copy. Elrod estimated a net of $750 for a printing of 5,000 copies.[167] Unfortunately, Mills died before Elrod completed the project, and he had to start anew with Agent O. J. McGillis.[168] Elrod sent the prospectus and samples of the draft text, explaining that he had secured more than two hundred photographs, many he himself had

taken and others for which he had permission.[169] He intended to repli-
cate the quality of his *Guide,* "the bible of the park for those who carry
it," to sell for about a dollar, depending on the quality of the cover.[170]

Elrod counted on numerous buyers of the book because of its appeal
as a souvenir. Once again, however, bad luck plagued him. McGillis
cited the downward economic trend that began with Black Thursday
in 1929 for not proceeding unless conditions changed. He left Elrod the
option to complete the project on his own or drop it.[171] As some solace,
McGillis ordered copies of the revised *Elrod's Guide* for the 1931 season,
having on hand enough for 1930. He also hinted at the railway's willingness to purchase and sell copies of the flower book if Elrod published it.
However, Elrod, too, declined in view of economic conditions.

He still found a way to profit from the park's plants and flowers, by
harvesting and selling seeds as he had done years earlier with tree seeds.
In 1927 he collected so many seeds that he ran out of envelopes with a
mass of seeds yet to package.[172] He estimated a total value of $130 for the
seeds representing a variety of species. He sold directly to people who
visited his exhibits and by mail order to those who learned of the service. Revealing his diligence and resourcefulness, he collected bear grass
seed at Grinnell Lake and Canyon Creek by using the Glacier Park Hotel
Company's Cadillac for the excursion.[173] Timing mattered for some species, such as bear grass and the glacier lily. He described sales as brisk,
noting on another occasion that the business continued to thrive. While
he never calculated overall total sales, Elrod clearly profited from this
project, incidental to his required work in the park.[174] He had envisioned
an even larger market for the flower book, but he settled for the proceeds
of the seed sales.

VIII

Much the same fate doomed his proposed history of the park, although
fragments of it lay strewn throughout Elrod's papers.[175] He initiated the
several pieces about the early pioneering visitors to the park region for
inclusion in the history.[176] In addition, he exalted the glaciers, little known
or studied before his years in the park. In 1926, Elrod corresponded with
George Bird Grinnell about Grinnell Glacier. When they met by chance
in the Many Glacier Hotel that same year, "It had been years since the

friends had seen each other, and . . . the conversation turned to days gone by." The following day, Elrod and Grinnell, with a large party, visited Grinnell Glacier and discussed the changes they observed.[177] Shortly after the visit, Grinnell related to Elrod his initial exploration of the glacier on a hunting trip in 1887.[178] According to Grinnell, Colonel John H. Beacom named the glacier when the two of them explored it that year. He sent to Elrod a letter from M. W. Beacom, the colonel's brother, documenting the visit and the naming to confirm the date.[179]

Requesting their return, Grinnell included photographs of the glacier taken in 1885, 1887, and 1888, noting that the photograph from 1885 showed "almost precisely the view . . . you took last July."[180] The photographs revealed clearly that the ice continued to the Continental Divide in 1885 but had receded significantly by 1926. Recognizing the deterioration, W. N. Noffsinger, Professor E. V. Huilington of Harvard, and Elrod had placed twelve stakes in the glacier in September 1922 to chart its movement and erosion.[181] Noffsinger later found that some of the stakes had fallen down and proposed resetting them. As he reported, "The top of the glacier seems to have melted in the meantime for six or eight inches." On the basis of these observations and discussions with Grinnell, Elrod documented the glacier's retreat of some two hundred feet by 1926.

In his annual reports, newspaper articles, and photographs, Elrod celebrated the glaciers, discussed their origins and development, and chronicled the changes he observed or deduced from the geological evidence.[182] The remains of an awesome icy world left high on the mountains, the glaciers became isolated centuries earlier as the continuous ice flowed inexorably to melt in the plains below. Over the ensuing centuries, the glaciers had advanced and receded as the result of new snow and changes in the climatic conditions, attaining their greatest magnitude most recently in about 1875 and had declined since then. He fully expected equally dramatic changes to continue until the glaciers dissipated completely. In 1931, he explained that Clements Glacier (also referred to as Museum or Two Oceans Glacier), greatly reduced in size from its glorious state when he had first seen it two decades earlier, got its original name because the water created by the melting ice went to the Pacific Ocean and the Gulf of Mexico.[183] State university president Charles H. Clapp, a geologist, used Clements Glacier to illustrate the natural processes of glacial activity. In 1925, Park Superintendent C. J. Kraebel marveled at the photographs Elrod included in his annual report that showed a receding Blackfoot Glacier.[184] At Kraebel's urging, Elrod prepared an article with photographs to illustrate the erosion process.[185]

That same year, Elrod returned to both Blackfoot Glacier, which once covered fifteen square miles, and Harrison Glacier, which he thought the "most wonderful and the most beautiful glacier" in the park, having first visited them in 1909.[186] He found the changes in Blackfoot Glacier overwhelming. "My first exclamation was 'Where is it?'" Instead of the huge pyramid of ice they had walked around and admired, they strolled over the bare rock floor, "as easily as one would an inclined pavement in Helena," for half an hour before they reached the ice. After fifty minutes crossing the ice field on the Continental Divide, they finally caught sight of Harrison Glacier, "nearly gone, bare rocks showing where I had seen such wonderful . . . blue crevasses . . . years ago."

On that same trip, the party traveled over very rocky terrain to Mount Logan.[187] "This mountain side was formerly covered, when the great glacier extended down the valley, moving slowly, scouring the sides of the mountain . . . and far out on the plain." Based on the evidence, he concluded that "great blocks have been falling for years, to melt at the foot of the cliff and be carried away to the Pacific." In fact, a huge ice slab fell during the visit. The evidence revealed the processes involved, leaving "the most awesome mess of broken rubble," with "water trickling everywhere, but almost no life." He wondered about the ancient trees on the moraine: "How many times have they died?"

Elrod's multiple photographs of the glaciers coupled with his observations to provide an early perspective on the future of these remnants of the glacial age. Years later, after perusing Elrod's reports and photographs, Park Superintendent J. W. Emmert deduced that most park glaciers had lost more than half their magnitude over a quarter of a century.[188] However, he also thought the decay had reached its nadir. To support his assessment, he noted that in 1951 the depth of Grinnell Glacier advanced from ten to twenty feet and the front rose some five feet. Elrod probably greeted with relief the superintendent's optimistic expectation of the persistence of the name Glacier National Park rather than "Glaciated National Park." However, time and other developments after Elrod's death in 1953 have proven the accuracy of his earlier and much less sanguine prediction for the glaciers.

As he explored and celebrated Glacier National Park's marvels, Elrod also indulged a lifelong fascination with mountain climbing and mountain climbers.[189] In an early article, he vividly described an electric storm in the Mission Mountains that evoked the acrid odor and taste of sulfur and forced him to toss his camera gear.[190] He climbed almost every peak in the Mission and Swan Ranges, and many in the park. As mentioned earlier, in 1906 he tried to get to the top of Mount Saint Nicholas but never made it.[191] His review of mountain climbing in the park began with Norman Clyde's conquest of thirty-seven peaks in 1923, an additional twenty-three in 1924, and more later.[192] In 1930, the Reverend Conrad Wellen of Havre sent Elrod an account of his successful ascent of Mount Saint Nicholas.[193] Wellen said he had learned of Saint Nicholas from Superintendent Eakin, who told him of peaks not yet climbed. Shortly thereafter, he saw the description in *Elrod's Guide* and developed a plan to conquer the mountain by approaching it from the juncture of Park Creek and the Middle Fork of the Flathead River. To commemorate the achievement, he built a cairn on the summit. Elrod included Wellen's accomplishment in his proposed book and also praised the achievements of Peter Hauser, an Austrian who climbed twelve peaks around Logan Pass in the late 1920s.[194]

However, he reserved his highest accolades for the climbing philosophy and accomplishments of G. M. Kilbourn, a walking guide for the Park Saddle Horse Company.[195] In his free time, Kilbourn climbed fifteen peaks, five of them twice, always using a different route for the second ascent. His climbing philosophy fascinated Elrod. Kilbourn described himself as "an incurable sightseer" with "a weakness for every upward slope and summit," who had never traveled to Switzerland, never owned hobnail boots, and learned all that he knew about "high-hung snow and ice routes" by climbing. Unconcerned about the predictable reaction of experts to his experiences, "I shall be indeed gratified if I can kindle in others . . . [an] unquestioning enthusiasm of eye and foot which makes every distant peak an almost irresistible challenge."[196] Kilbourn certainly found a kindred spirit in Elrod.

Elrod also described in detail some of the interesting people he met in the park. On one of his excursions, he spent a few hours with Arthur

Reynolds, an exemplar of the experienced and independent woodsmen rapidly disappearing from the West and from the park.[197] Although he chose to live in the wilderness, Reynolds sent his son to Harvard. Employed by the Forest Service for nine years during his younger years, he worked as a guard for the Park Service and lived year-round in the park when Elrod met him. Then sixty-four, he eschewed tobacco in any form, exuded health and vitality, enjoyed company, willingly shared his dinner, and walked fifteen to thirty miles every day, rarely carrying a lunch (to remind himself to return home). Armed only with a small hatchet to cut trees and brush out of his way, he found it adequate to deal with any challenge.

Reynolds became famous for the signs he posted as he walked the mountains; they boldly proclaimed "Death on the Trail." When asked about them, he said he enjoyed the reactions of impressionable tourists who turned back for fear of actually encountering dead men. Elrod espe- cially enjoyed an amusing incident Reynolds related concerning a young Englishman on his first visit to the park. He took Reynolds's advice to buy three-foot snowshoes rather than hobnails to hike in the park. How- ever, Elrod reserved special respect for park guides like Kilbourn, the avid mountain climber, and others who found the naiveté of the tourists irresistible. One grizzled old guide, Jack Brown, who had probably heard the question once too often, responded to a woman who asked about guide duties in the winter with the quip: "Lady, the winter is so long, the snow is so deep, and it drifts badly. It keeps the guides busy sweeping off the passes so they may be ready for the tourists in the spring."[198]

Other stories ranged from the humorous to the tragic. On a horseback ride to Iceberg Lake, one portly woman used pillows in front and back of her to soften the saddle. Another woman, confused by the width of the saddles, specifically asked for thin horses. The wrangler advised her to return later in the season when the horses had lost weight because of the hard work of carrying tourists.[199] A woman ventured into the rocks above Lake Grinnell, got stuck, and the rangers had to rescue her. In 1925, two children fell off the boat on Swiftcurrent Lake and drowned. Even experienced mountaineers encountered difficulties: one died of a heart attack trying to climb Mount Siyeh.

Eakin told Elrod about the plight of two men, two women, and a boy who tried to walk from Sperry Chalet to Granite Park, even after the ranger warned them against it. The Park Service considered a trail through the area too dangerous, and Elrod thought only experienced mountaineers had any chance of traversing it. One of the men slipped and fell thirty feet into a crevasse above Twin Lakes. The other man and the boy lacked the strength to get back after going partway down into the crevasse in a futile attempt to rescue their companion. Rather than leave immediately, the two women waited twenty-four hours before walking back to Sperry Chalet for help. One of the women told Elrod later that she had read his *Guide* cover to cover six times while they waited. Too many tourists thought they had only to know the trails, and forgot that they also had to recognize the perils. The man who fell subsequently died and it took four full days for twelve men to carry his body out through underbrush too dense for horses. Elrod mused to Emma that "Glacier Park is no place for fools, careless persons, and egotists who want to do stunts, or children. No one can play with the mountains, for they are sober all the time. But I do love them very much."

Elrod's fascination extended to animal life in Glacier National Park as well. In "Conservation," a piece he wrote during the mid-1920s, he emphasized protection for the increasingly challenged flora and fauna of Montana.[200] Years earlier he had warned of the increasing scarcity of native birds and animals. He dreaded their extinction, with the buffalo gone and the grizzly bear confined in an ever smaller, more remote space and hunted mercilessly.[201] The antelope had little chance of survival because of polluted rivers, intrusive fences, and uncontrolled slaughter. Calling for balance, he urged national legislation to limit state action and individual behavior. In that regard, he believed that national parks, especially Glacier National Park, had the potential to further preservation and conservation.

The census of large animals compiled by the Park Service between 1921 and 1924 corroborated the promise. Animal numbers had increased substantially to a count of 4,505 (with an estimated total of 6,384), including 51 (104) silvertip grizzlies, 76 (148) black and brown bears, 724 (1,111) mountain sheep, 943 (1,600) mountain goats, 69 (88) moose, 567 (706) elk, and 2,056 (2,626) deer.[202] Although no baseline numbers existed, Park Superintendent Kraebel hypothesized that game animals had increased at a steady pace. Elrod extrapolated from the 1924 data, using the growth rate of the National Bison Herd in Ravalli since 1910, and predicted a count of 71,950 (estimated total of 103,104) animals by 1936.[203] Conceding much slower growth, he lowered the multiplier to five and quite accurately predicted an animal population of roughly 30,000 by 1936.

However, he denied the park's safe harbor to predatory animals such as the wolverine, coyote, wolf, mountain lion, and the Canada lynx. He had found the wolverine rare, the wolf even rarer, the lynx nearly gone, the mountain lion checked, and the coyote the object of open war.[204] In another piece about the coyote, he reported without criticism that park employee Chance Beebe killed fifty-four coyotes in two months, nearly one a day.[205] The naturalist's concern for conservation and preservation—generous to a fault to all flora, other fauna, glaciers, and mountain men—did not apply to predatory animals.

IX

Elrod's summer seasons in the park, with a small salary, room and board, and travel expenses, afforded wonderful opportunities for research, writing, and escape from mundane demands. However, he lost income when in 1924 the reorganized Montana Education Association launched its official journal and, after public, less than credible, and hostile criticism of Elrod for profiteering, forced an untimely end to the *Inter-Mountain Educator* he had edited and published for a decade.[206] He assured Emma that even if he lost the journal, other activities offered better returns with less work and worry.

Despite his nonchalance, he suffered an ulcer attack as a result of the stress of the controversy over the paper. His brother-in-law, John Hartshorn, urged Elrod to check into the Mayo Clinic in Rochester, Minnesota.[207] Minimize the risks, Hartshorn counseled, and he offered to pay all expenses for the trip and medical treatment. Reinforcing the message, he sent a telegram in which he advised Emma to persuade Elrod to accept the offer. As he counseled, Mayo Clinic physicians operated only as a last resort.[208] He wrote again to emphasize the seriousness of the affliction and his willingness to cover all expenses.[209] Nonetheless, Elrod remained in Missoula under the care of his local doctor, and proper diet and rest healed the ulcer. Relieved by the rapid improvement, Hartshorn counseled Elrod to follow the diet, stay away from work, and at all costs avoid an operation more likely to exacerbate rather than cure the malady.[210]

Elrod followed the advice about diet and work and returned for the 1925 season in the park, determined to rest, eat the right food, avoid mountain climbing, and limit his work hours.[211] He assured Emma that Many Glacier Hotel provided healthy fare such as poached rather than scrambled eggs, fish, vegetables, and plenty of milk.[212] He felt much better, irritated only by the disgusting jazz demanded by the tourists that disrupted his evenings. As a helpful gesture, Hartshorn bought a used Cadillac sedan with 22,000 miles, had it completely refurbished, and sent it prepaid by rail to Butte for the Elrods, the first car they ever owned.[213] Adding new worry, however, Elrod's son-in-law suffered severe injuries from an airplane accident that summer.[214] Elrod feared that Billy and Mary faced some tough times, and he urged Emma to take steps to avoid any financial distress for them. As it happened, both Elrod and Billy recovered fully, the Cadillac arrived in Missoula without incident, and Elrod found other ways to sustain the family income.

Despite the financial setback, Elrod's position as a seasonal park naturalist continued to improve. In February 1923, Eakin and Noble accepted his proposal for a modest expansion of the Nature Guide Service.[215] In May of that year, he planned more exhibits and lectures and requested three temporary naturalists to assist him.[216] One of the assistants, a married man, wanted to bring his family with him but requested a tent and other necessities. Elrod prepared detailed descriptions of the duties, secured approval of the arrangements for the married assistant, and asked specifically for guidance about the services at Lewis's private Glacier Hotel on Lake McDonald. He frankly thought Lewis interested only in the money, with neither knowledge of nor concern for the park and the public interest. Given Elrod's own keen eye out for lucrative opportunities, the criticism sounded harsh even if accurate. Nonetheless, the services remained the same at all locations. As for research in the park, he informed President Clapp that both Eakin and Noble offered to support relevant research projects at Many Glacier.[217] To that end, he discussed a possible building with Eakin, and asked the Glacier Park Hotel Company for the use of a chalet. Outlining ambitious plans, he suggested extending the collaborative project to Yellowstone Park. Nothing came of these possibilities, however, except a trip to Yellowstone.

Over the next few years, Elrod guided the development of the educational policy of the National Park Service and defined the duties of ranger naturalists under the supervision of the chief park naturalist.[218] Elrod's Park Naturalist Service reports for 1925, 1926, and 1927 set out an ambitious agenda of activities and recommendations.[219] Specifically, he urged an educational program for guides and bus drivers to inform them about park features and history. He also recommended expanded tours of the glaciers with guides stationed at some glaciers, nightly lectures followed by movies, and the construction of new trails for the benefit of tourists. In addition, he wanted improved exhibits and signage and an enhanced sales program for government publications (including *Elrod's Guide*). In one communication, he recommended twenty-nine ranger naturalists during the summers because of projected increases in tourism and expanded services. Most of his aggressive proposals unfortunately failed to win approval.

In July 1926 Superintendent Eakin and Dr. Ansel Hall, the new chief park naturalist, came to the Many Glacier Hotel to discuss the educational program in the park. Elrod thought the discussion went well.[220] His evening talks at the hotel for the tourists proved increasingly popular, and he introduced colored slides and movies to enliven the lectures.[221] He also discussed with W. H. Mills, general advertising agent for the Great Northern Railway, a two-week educational summer session for tourists.[222] Although nothing came of that discussion because of Elrod's insistence on instruction of the highest quality in appropriate facilities, Elrod continued to press for more comprehensive educational services.

As early as 1923, he proposed an appropriate educational facility and museum at Many Glacier.[223] Superintendent Kraebel subsequently decided to place the main park museum in Belton and to make the complete exhibit of park information and materials accessible year-round there.[224] He thought a relatively small but fireproof structure adequate for a few birds, flowers, plants, small animals, and minerals at Many Glacier, eliminating the work spaces and large animal specimens that Elrod envisioned. Elrod disagreed and never changed his mind about the location of the main park museum. To his chagrin, instead of the twenty-nine assistants he proposed—or at least a significant increase— Elrod reviewed applications for temporary ranger naturalists sent by

Superintendent Kraebel, who indicated an intention to recommend at least one and perhaps two.[225]

By 1926, Elrod had the formal title of park naturalist at a prorated summer salary, and by 1928, his prorated salary had increased from $100 a month in 1922 to $145.[226] Several letters to Emma included detailed descriptions of his duties, from traveling around to visit his assistants, helping them to construct exhibits, preparing press releases, and interacting with visiting dignitaries.[227] The park superintendent apologized that regulations prohibited the acquisition of books for a library, but he approved Elrod's requests for upgraded displays and photograph enlargements. Elrod's supervisors also encouraged his visits to various sites around the park to monitor the responsiveness and quality of the services. In 1928 Elrod mentioned with relief the promise of another assistant to assume the routine tasks and allow him to devote more attention to his administrative responsibilities. In August 1928 he received assignments from Eakin and Hall to prepare formal reports, indicating further confidence in and expansion of his official duties. In 1929, he eagerly accepted the invitation to return to the park with the same salary and two assistants.[228]

With characteristic energy regardless of his assigned responsibilities, Elrod attempted to establish the Glacier National Park Museum. George Bird Grinnell consented to serve as honorary president and Elrod became president of the Associated Museum Society.[229] He wrote his plan for the museum and included a schematic of the facility (60 feet by 120 feet) that showed a great room (45 feet by 90 feet) for public meetings, workrooms, and laboratories, and spaces for storage, an herbarium, exhibits, and a photography darkroom.[230] He designed the facility to serve the general public, amateur naturalists, college students, and professionals. He also envisioned floral displays changed daily. With Kraebel's departure and Eakin's return as superintendent, Many Glacier became the preferred site once again. Elrod obviously believed he had a chance of realizing his educational vision for Glacier National Park.

However, once again reality rudely disrupted his plans. In 1930, he learned the fate of the flower book and other effects of the economic depression that had struck in 1929.[231] Noffsinger complained of $9,000 less in gross revenue than anticipated, resulting in a loss for the Saddle Horse Company that year.[232] Looking for someone to blame, Noffsinger criticized the new chief park naturalist, George C. Ruhle. As he remarked snidely, the chief naturalist had no practical business sense and failed in his duty to promote tourism.

For his part, Ruhle invited Elrod to return for the 1930 season with the same duties plus lecturing assistance by two visiting professors from Wisconsin and Harvard.[233] At the same time, he broke the distressing news of a revised agreement with the Glacier Park Hotel Company that eliminated free room and subsistence and required the temporary naturalists to eat with the bus drivers at forty cents a meal and live in tents. In partial compensation for these dramatic departures from life in the Many Glacier Hotel, the proposed salary increased modestly to $155 per month. Elrod declined the invitation, citing his responsibility to direct the research sponsored by the Fish and Game Commission at the Biological Station at Flathead Lake that had actually begun in 1928. Without question, the changing conditions and relationships weighed on his mind as well. The alluring joint venture of 1922 seemed about to become a part-time job.

As if to confirm Elrod's suspicion, one of the temporary ranger naturalists complained to Ruhle, with a copy to Elrod, that the company had barred him and others from the hotel dances and required them to take *every* meal *only* in the transportation mess with the bus drivers. The new hotel manager told the temporary ranger bluntly that he considered the naturalists equivalent to waiters and "gear jammers," and he intended to treat them the same.[234] In May 1930, Eakin thanked Elrod for his service and expressed regret about his decision to forego summers in Glacier National Park. However, he understood the reason under the circumstances.[235]

Eakin also informed Elrod that the Park Service had acquired the Apgar property at the foot of Lake McDonald and planned to remove the private buildings and reforest the entire area. In addition, he had received funding for a full force in 1931, including the park naturalist and six temporary ranger naturalists.[236] The economic conditions continued to worsen, with rail traffic off by some twenty percent. Eakin hoped for improvement in automobile traffic with the completion of U.S. Highway 2, the Roosevelt Highway. However, by the time he received the letter, Elrod had returned to his first love and resumed leadership of the Biological Station for another new beginning in 1928.[237]

Elrod relished his years as a naturalist in Glacier National Park, although he occasionally complained about the drudgery. The Park Naturalist Service he developed enhanced the park experience for all who came. *Elrod's Guide* informed people around the world and allowed them to appreciate nature's wonderland in Glacier National Park. In the draft introduction to the planned but unfinished history of the park, he wrote proudly of his early expeditions, scientific studies, and hundreds of photographs. Over the years, he had produced colored slides to illuminate his lectures about the wonders of Glacier National Park that he delivered in and out of Montana. He intended to include the lectures and slides in a book and hoped to reach a much larger audience. He urged prospective readers who embarked on a real or virtual visit to special places such as Glacier National Park, to learn as much as possible about the place in advance of the trip, and to plan related study and research along the way. Focus and planning enhanced the ultimate pleasure of reaching the desired destination. Elrod thought that lesson applied to all of life: "For such work the best thing is to ride a hobby."[238] Without question, he had followed his own advice, much to the great enjoyment and benefit of thousands.

CHAPTER FIVE

Reprise: The Biological Station
and the Golden Past

Even before the Depression struck and ended Elrod's Glacier National Park interlude, he had succumbed once again to his obsession with the Biological Station at Flathead Lake. In 1926, he abruptly rejected an offer to buy station land around Yellow Bay that had been reserved by Congress for biological research.[1] Shortly thereafter, when Dean C. W. Leaphart of the School of Law objected to his annual request for station funding, Elrod reminded President Charles H. Clapp that for years the station had attracted eminent scientists from across the country to do research.[2]

His response put everyone on notice that he still cared deeply about the station and intended to protect it. Late opening decisions in prior years had caused low attendance of students and researchers. Despite the fact that sustaining the station required constant and ongoing effort, he found the labor well worth the time and energy. In fact, he considered the work at the station far more important to the university and science than the project in Glacier National Park or most other activities at the university. He literally commanded the president to inform the chancellor and anyone else of his desire to reopen the station.

I

With these comments, Elrod passionately expressed his yearning for the Biological Station to prove its unrealized potential. He made the comments partly in response to a 1928 message from Chancellor Melvin A. Brannon about some research the State Fish and Game Commission proposed, offering opportunities for the university's science departments and the station.[3] Brannon came to the university as its second chancellor—when Elliot left for Purdue in 1922—after a successful tenure as president of Beloit College.[4] A biological scientist, Brannon identified closely with the State University and served as chair of the committee that petitioned for a Phi Beta Kappa chapter in 1929. His term as chancellor ended with his resignation in 1933 when disgruntled members of the state legislature voted to abolish the office of chancellor in retaliation for his independence and defense of university prerogatives.[5]

At Brannon's suggestion, the Fish and Game Commission unanimously authorized its chairman, Thomas N. Marlowe, to confer with Elrod to develop a work plan and budget for the new Biological Station project. Although contracted for his summer work in Glacier National Park for 1928 and 1929, and forewarned by his experience of a decade earlier, Elrod welcomed the opportunity. Even so, the chancellor complained in May about Elrod's tardiness in scheduling the discussions, urging President Clapp to monitor Elrod closely and insist upon prompt action.[6] Clapp immediately established a research council that included himself and the chancellor to keep Elrod on track. Oblivious to these developments, Elrod proceeded at his own pace, intent upon resurrecting the Biological Station as the leading center for the investigation of Montana's natural history and resources.

After some delay, Elrod and Marlowe agreed in May 1928 on a research project focused on the fish and fish food in Flathead Lake, similar to the one Elrod had conducted a decade earlier. This time, however, the budget of $4,000 for each of two years augured real progress toward identifying the species, estimating the numbers of fish in the lake, and establishing the lake's capacity to produce fish.[7] In the planning discussions, Elrod summarized what he already knew about Flathead Lake and its fish, and identified three specific objectives for the project: (1) the availability of food and breeding grounds for fish; (2) the species, numbers, distribution, migration patterns, habits, and interactions of the fish; and (3) the maximum carrying capacity of the lake.[8] With an eye toward the future, he also suggested some related studies and listed the relevant station publications and reports on the fish and birds of the region. Because of his obligation for two more summers in Glacier National Park, he agreed to serve as the principal investigator in absentia, without pay, and to supervise assistant station director Robert T. Young's fieldwork at the station.[9]

The collaborative nature of the new project boded well for its success. President Clapp contracted for a geological study of the lake, while Elrod committed to summarize and supplement his earlier work on lake water levels, evaporation, and vertebrates. Assistant director Young studied the microscopic plant and animal life of the lake, physics professor G. D. Shallenberger accepted the challenge of establishing light penetration and water currents, and chemistry professor J. W. Howard identified several relevant chemical analyses. Botany professor Joseph E. Kirkwood proposed to characterize the botany of the region, a study continued by Professor Charles Walter after Kirkwood's untimely death in 1928.[10] The chancellor agreed to coordinate the billing and reporting processes. As the final step, Elrod and Marlowe set up a steering committee consisting of Elrod, Young, Chancellor Brannon, President Clapp, Chairman Marlowe, and I. H. Treece of the Fish and Game Commission.[11]

Elrod visited the Biological Station once in 1928 with Marlowe, Brannon, and Clapp to review progress.[12] He found the station in good repair and the project functioning well. The group went about two miles out on Flathead Lake to visit a floating platform designed for taking readings and samples of water and mud. For his part, Marlowe expressed satisfaction and commended the progress. As it turned out, Marlowe accepted the preliminary report, approved its publication in the commission's annual report, and authorized continuance of the project for another year.[13]

Elrod and Marlowe outlined the plan for 1929, essentially the same as 1928 but with the addition of some bacterial studies and work on Georgetown Lake, a critical source of eggs for planting by the state hatcheries.[14] Victor E. Graham, who had worked on Wisconsin lakes, joined the team to analyze bacteria in the lake, while Young and Graham looked for any diseases or parasites affecting the fish in Georgetown Lake. Elrod also suggested consideration of a future project to study ducks and fish in eastern Montana lakes, with the possibility of discovering new techniques or species for introduction to the Flathead Lake region.

Prior to the final conference in 1930 on the two-year project, Marlowe forwarded to Elrod a letter from E. L. Wickliff, chief of the Bureau of

Scientific Research in the Ohio Department of Agriculture, with details about planting whitefish eggs in Lake Erie.[15] At Marlowe's invitation, Wickliff planned to visit Montana to evaluate planting Lake Superior whitefish eggs in Flathead Lake. To facilitate the visit, Marlowe directed Elrod to send Wickliff the data about the lake from the 1928 and 1929 studies.[16] In the same letter, he requested analyses of the Clark Fork River, increasingly discolored by some unknown cause. Since the commission had stocked the Clark Fork with trout, Marlowe wanted advice about ways to protect this prime fishing river. To discuss these and other matters, Chancellor Brannon charged Elrod to attend the meeting of the Fish and Game Commission in May 1930 to report on the achievements of the project in 1928 and 1929. Brannon also instructed Elrod to answer any budget questions, request payment of the entire $8,000 budget for the two years including the funds not expended in 1928, and present snappy requests for future projects at Flathead Lake and elsewhere.[17] Given Marlowe's instruction to send data to Wickliff and his request for some work on the Clark Fork River, and considering Brannon's instructions to request funding for more projects, Elrod had every reason to be optimistic about this new beginning for the station.

Chancellor Brannon's introduction to the report celebrated the scientific collaboration of government and University of Montana scientists. The report itself summarized the findings of the investigators and incorporated data from Elrod's earlier reports.[18] On the basis of 127 soundings, the deepest to 329 feet, Shallenberger determined that the sun's illumination penetrated to 140 feet in foot-candles. The team further determined that water at the lake bottom had an average temperature of 39.6 degrees Fahrenheit, and that the lake provided sufficient food and oxygen for fish life at all depths. In the lake proper, Young identified 130 species of microscopic plants and animals distributed among 81 genera, including blue-green algae, diatoms (phytoplankton), desmids (green algae), and flagellates (protozoa), numbering on average to 58,200,000 organisms per cubic meter of water.

On the basis of those data, they judged life in Flathead Lake sparse compared with other lakes. The marshy areas around the lake edges proved much more productive, though still less so than similar areas elsewhere. Young and his assistant captured specimens of roughly one hundred species of several million microscopic organisms per cubic centimeter, still less than other more productive lakes. In brief, as Young concluded in his study—now regarded as a classic—of the microscopic plant and animal life in Flathead Lake, the research revealed a deep, cold lake with low productivity, not particularly promising for fish culture.[19] Graham's study of lake bacteria identified nothing of significance, and the evaluation of Georgetown Lake found no diseases or parasites threatening this source of fish eggs for the commission. Walter completed Kirkwood's inventory of botanical resources, and Howard found nothing abnormal in the lake chemistry.

Elrod summarized his prior analyses of nearly five hundred fish stomachs to identify the food preferences of the various fish species. In addition, he drew on his studies of the lake's rise and fall, evaporation, and other features, and consolidated several earlier reports on fish species and their relative abundance in the lake with new data discovered during 1928 and 1929. The resulting list of the species included common suckers, long-nosed suckers, squawfish, Columbia River chub, bull trout (by various names), native trout (by various names), Rocky Mountain bullheads (by various names), native whitefish (by various names), Lake Superior whitefish, Mackinaw or lake trout (by various names), Chinook salmon (by various names), sockeye salmon (by various names), eastern brook trout, rainbow trout, black bass, smallmouth bass, bluegill sunfish, and bullheads. To put it all together, Elrod submitted seven introductory pages, five pages of recommendations, twenty-one photographs, a copy of President Clapp's geological map of the lake and surrounding area, the outline of a plan to deal with the Clark Fork problem Marlowe had identified, and a proposal to study certain lakes in Glacier National Park.[20]

Elrod's summary informed the commission that successful fish culture in the lake depended on the implementation of five specific strategies.[21] First, systematically increase the native cutthroat trout, since the lake had once supported a superb trout fishery, and he thought this fish well suited to its ecology. Second, cultivate the native whitefish, a good food and sport fish which had always done well in the lake and its tributaries. Third, make further attempts to plant Lake Superior whitefish, since recent samplings indicated the fish had begun to adapt and propagate, although the carnivorous lake trout, bass, and other species had severely reduced its numbers. Fourth, make careful use of tributary lakes and streams for plantings and migration to Flathead Lake, specifically Swan, Bowman, McDonald (Mission Mountains), and others. Fifth, and most important, reduce by all possible means the predators in the lake, particularly the lake trout, bull trout, suckers, squawfish, and black bass.

As the most effective ways to assure the elimination or at least drastic reduction of the predators, Elrod recommended (1) opening the lake to commercial fishing for designated species, (2) allowing the public sale of bull trout taken in the lake, and (3) placing a bounty on suckers and squawfish. In the end, however, perhaps recalling his earlier experience, he emphasized the caveat that successful sport or commercial fishing in Flathead Lake depended on rigorous control of predators, repetitive plantings of desired fish, and systematic monitoring. In essence, he followed Brannon's instructions and presented an impressive and thorough report that laid the foundation for future studies by the Biological Station at Flathead Lake. The way appeared open for future work in accordance with a comprehensive plan to increase the production of desirable fish in Flathead Lake.

II

Despite Elrod's best effort, the Fish and Game Commission funded no more station projects, in all likelihood because of the worsening economic conditions in the state and nation. In his final report on the project, Elrod carefully noted that he had netted some two dozen descendants of the Lake Superior whitefish planted years earlier, demonstrating that the fish had indeed survived. He also caught a few sockeye salmon, auguring success for that fish.[22] Years later, a researcher at the Biological Station at Flathead Lake confirmed Elrod's warning about predators. Bonnie Ellis, Research Assistant Professor of Limnology, concluded that the lake trout—planted a century earlier but held in check by the lack of a food source until the deliberate introduction of mysid shrimp in 1968 to 1978 as food for kokanee salmon—ultimately and dramatically "flourished on mysids and decimated the kokanee fishery."[23] Elrod's final recommendations had hit the mark squarely, and the predators won the struggle for survival. However, despite these longer term and as yet unanticipated consequences, nothing at the time resulted for the station.

Hoping to build on the successful project, Chancellor Brannon proposed in 1930 a collaborative study of Glacier National Park lakes with the Federal Fish Hatchery but received no response.[24] Elrod published an article on the Montana grayling at the invitation of Floyd Smith, editor of the Fish and Game Commission's journal, and defended the commission's efforts to propagate that fish.[25] In 1931, a news article reported the planting of 300,000 Lake Superior whitefish in Skidoo Bay on the southeastern shore of Flathead Lake, a positive result of Elrod's recommendations.[26] Elrod certainly had reason for satisfaction with this vindication of his earlier prediction that fish adapt, propagate, and migrate, given time.

Even so, no more projects materialized for the station, perhaps partially because of competing expertise in the state. In 1929, Elrod asked William Ferguson to inquire about Hauser Lake and other lakes around Helena.[27] Only a discouraging response came from State Game Warden Robert H. Hill, reporting that Hauser Lake had little to offer because of cloudy water and carp infestation.[28] In 1931, W. J. Thompson, superintendent of the Bureau of Fisheries in Bozeman, told Elrod he had no funds for Glacier National Park lake studies, but he referred Elrod's inquiry to experts in the Bureau of Fisheries.[29] More than a year later, A. S. Hazard, associate aquatic biologist of the Bureau of Fisheries, explained to Elrod his two-year project to study fish culture in the park's streams and lakes, using careful soundings, evaluation of food sources, and analyses of fish stomachs and health. He also informed Elrod that he had all the technical experts the project required, but he did enclose a check for a copy of *Elrod's Guide*.[30] In contrast to his earlier experience, the aging naturalist no longer enjoyed the advantage of superior expertise in his field. Time once again had altered the conditions and shifted the balance against him.

An auspicious occurrence in 1929 revealed how passing time, inadequate information, and faulty communication can wreak unanticipated injury. With Elrod present, the State University Campus Development Committee voted unanimously to designate the entire campus as the Elrod Bird Sanctuary, with plans to include a birdbath, nesting thickets, berrying shrubs, and feeding shelters at the foot of Mount Sentinel.[31] *Montana Wildlife* featured an article on the occasion, praising Elrod's distinguished service, specifically the fish study of 1928–1929 at the Biological Station on Flathead Lake "operated by the State Fish and Game Commission."[32] Despite his labor of love for three decades, the private support he had personally secured, and the battles he had waged to protect the station, he suffered the ignominy of seeing the station described as an arm of the Montana Fish and Game Commission, the agency he held largely responsible for the station's closure from 1921 to 1928.

An even stronger sense of frustration must have claimed him when Frank A. Nolan proposed the transfer of station land at Yellow Bay to Montana State College at Bozeman for use as a picnic ground and demonstration cherry plot.[33] Highly indignant, Elrod vented his outrage at this proposed violation of the terms of the congressional grant. He sharply denied that the current lack of funding for the station preempted future possibilities. In any event, he dismissed the absurd idea of giving priority to entertainmnent, commercial, or agricultural schemes for the use of the only remaining research site on 150 miles of lake shoreline. After all, he had set aside some land for a public picnic site years earlier and saw no need for more.

He also reminded the president and the chancellor that he had invited the college to collaborate on the station as a joint venture in 1898, and when it declined the invitation, he took on the challenge by himself. Even more to the point, Senator Dixon had persuaded Congress in 1905 to grant the land for biological science purposes, and the university had accepted the grant as a contractual obligation. Prudence, federal law, and good faith had established an inviolable moral and legal trust. Agreeing with Elrod, President Clapp informed Chancellor Brannon that the Montana statute of 3 March 1905 accepting the grant for the Biological Station, as restated on page 25 of The University of Montana Code, did not permit either the state or the board to transfer the land to another institution or to assign a different use for it. Elrod heard no more of the proposal.

The issue resolved, Elrod submitted his last Biological Station report in 1931.[34] He and campus engineer T. G. Swearingen evaluated the facilities and equipment and compiled an extensive list of needed repairs, replacements, and maintenance. Nothing came of these recommendations because of funding constraints, and the station remained closed for sixteen more years, tended by a custodian.[35] Thus, while secure legally if not institutionally, Elrod's Biological Station did not open again until 1947, long after he had retired from active scientific work.[36] His last act for the station left the way open, however farfetched it seemed at the time, for yet another new beginning. Confirmation of his faith came in the ranking of his station as one of the top two in the world by the end of the twentieth century.

III

With Glacier National Park mired in economic doldrums, the terms of Elrod's service unfavorably altered for a man in his late sixties, and the closing of the station in 1931, he turned his attention once again to his family and the university. The healthy enrollments predictable during an economic downturn also engendered a higher failure rate, although he commended some first-rate students as well.[37] More depressing, reduced university funding resulted in fewer but larger classes and made teaching more stressful, accelerating the damaging trend he had foreseen years earlier.[38] Even more disappointing, as a former student lamented, the university had eliminated all science laboratories for freshmen.[39] Elrod believed that students learned only by using the scientific method in laboratories and on field trips, and that science labs offered the only defense against curricular triviality. The decision to eliminate them for freshmen must have come as a terrible blow.

In 1929 Elrod's daughter Mary divorced her husband of a decade and returned to live with her parents in Missoula.[40] Over the next few years, Elrod sought to help her secure professional employment, initially as secretary to the dean of women. Budget problems forced her to leave in 1933.[41] With her father's encouragement, she enrolled in Columbia University's Personnel Administration program and earned a master's degree in 1934. Elrod counseled about possible placements at Purdue, Stanford, or Northern Iowa, advising her to delay her choice until she heard from all three and to stress her executive skills and experience with young people.[42] When nothing came of that search, he asked President Clapp and the dean of women to help with positions open at Northern Montana College in Havre and the University of Arizona.[43] Whatever the outcome, even if all failed, "we are still not the worst off people in the world," Elrod assured his daughter. As it happened, she did not find a position, given the dire economic conditions for higher education in 1934, and a family tragedy forced her return to Missoula.

The impact of the Great Depression threatened institutional survival and mandated radical changes in functioning for higher education in Montana and across the country. Whereas the State University had enrolled 504 students and educated them in a physical plant of 6 buildings with a faculty and staff of 56 and a budget of $230,000 in 1915–1916, by 1933–1934 enrollment exceeded 1,500, buildings had grown to 13, and faculty and staff expanded to 92, but the budget had stagnated at $235,000.[44] In partial response, the university implemented salary reductions of nearly 20 percent between 1932 and 1934.[45]

Elrod continued his service as chair of several of the twenty standing and five special university committees, including the committee on budget and university policy, the campus development committee, the museum committee, and the committee on service.[46] In 1933, with Elrod's leadership and at President Clapp's request, the university committee on budget and university policy recommended the following:

- Denial of replacements for faculty members on leave;
- Reversion of all salaries released by vacancies;
- Reduction of all expense budgets;
- Prohibition on employment of faculty spouses or of married women with husbands;
- Retention of existing faculty and staff by reducing all salaries;
- Minimum salaries for instructors and janitors;
- Discontinuation of intercollegiate athletics, except contracted football games for fall 1933;
- Disallowance of fee deferrals for students; and, finally,
- A textbook fee for students, on departmental request.[47]

These drastic measures fully destroyed faculty morale and set the stage for even more chaotic turmoil on campus.

To deal with the crisis, President Clapp secured unanimous approval of a new curriculum, academic structure, and organization of the university.[48] The unanimous vote of the university curriculum committee opened the way for the segregation of the university into two colleges: a junior college for freshmen and sophomore students in four divisions, offering a spartan curriculum of interdisciplinary core courses, and a senior college for all juniors and seniors organized by majors and offering carefully sequenced advanced courses.[49] The students had to pass written or oral comprehensive examinations for admission to the senior college and major for which they qualified or leave the university. The senior college reshuffled the students by preference into majors managed by each department or school, with each major imposing minimal curricular requirements and the students free to choose other available coursework. The students who satisfied the minimal requirements received a pass and those who performed well on written or oral comprehensive examinations received honors at graduation. Clapp emphasized that the plan depended for success on faculty and student acceptance of accountability and responsibility for the quality of the education.

Under the proposed structure, the faculties worked directly with the students to monitor progress and certify the graduates, with the registrar and other administrators confined to record-keeping. As it turned out, the full proposal ultimately failed because the faculty "was not ready for such a change," as English professor and chairman of the Division of Humanities, H. G. Merriam, a participant, explained.[50] However, the divisional organization of departments in the College of Arts and Sciences persisted at least partially until the 1960s. The larger goals that Clapp, Elrod, Merriam, and others had in mind proved far too ambitious under the conditions at the time. Even before he suffered a paralytic stroke in 1934, Elrod had changed his mind about the merits of the new model.

Until incapacitated by the stroke, he labored to sustain the university by taking full advantage of federal stimulus funding.[51] Even minor campus beautification projects made a difference in the lives of faculty and students. Federal funds to provide work-study stipends for students also made it possible to sustain enrollments. A successful court challenge verified the authority of the state board to issue bonds of its own without binding the State of Montana. Thereafter, the university campuses qualified for federal loans and grants for new construction including, at the State University, a student union, journalism building, and chemistry-pharmacy building, all completed after Elrod's retirement.[52]

Elrod also managed tasks assigned to him or suggested by President Clapp. For example, at the president's suggestion, G. A. Jordan invited Elrod to assist in the development of Montana's phosphate deposits.[53]

Phosphate became increasingly important in the twentieth century as a component of fertilizer when mixed with sulfuric acid. Elrod declined to participate, resisting Jordan's plea to help find a new source of revenue for the state. The appeal failed to persuade him, perhaps because of the indelible memories of his earlier involvement as consultant for hire in the litigation against the Anaconda Copper Company in 1905–1906. In any event, he chose to occupy himself with more rewarding activities.

IV

Elrod's Charter Day address of 1916 (revised in the early 1930s but never again delivered), the Charter Day plans he disseminated during the 1920s, and his Aber Day address of 1932 revealed a great deal about his evolving state of mind.[54] In these speeches and plans, Elrod memorialized the growth and development of the university. To Professor William Aber, the father of the Aber Day tradition at the university, and J. H. T. Ryman, the longtime member of the local executive board and the state board, he gave plaudits for creating and maintaining the beautiful campus. The university had flourished under difficult conditions because of the dedication of faculty and students who from the outset worked harmoniously together. Every year after the university moved in 1898–1899 to the foot of Mount Sentinel, the campus community had come together on Aber Day to plant trees and clear away debris to preserve the legacy of the past.

Different in detail but similar in tone, Elrod's Charter Day speeches celebrated the fact that every faculty member and student had traditionally accepted the duty to participate in the annual birthday celebration that engaged the entire university and its supporting community. He outlined and promulgated the plans for the occasion in 1920 and reused them during the following decade. On behalf of the campus, he invited the public to a convocation with distinguished speakers, visitations to classes, tours of facilities and sites for new ones, student plays or athletic performances, a dinner preceded by a grand reception, and dancing and entertainment into the evening. In the 1930s, he passionately urged a return to the practices of old.[55] Aber Day and Charter Day held great significance because they commemorated campus traditions that symbolized the shared sacrifices and triumphs of those who built the university.

In reverential word and phrase, he exalted the contributions of dedicated administrators and faculty members over the years.[56] Aber set the mold by focusing on students and helping them to realize their potential.[57] Ryman, who on his death in 1926 left his estate to benefit students, contributed to the Aber mold with generosity and organizational skill during troubled times.[58] Together they bequeathed a rich legacy of shared commitment to productive engagement and celebration of accomplishment. Entranced by memories of days gone by, Elrod eulogized the ethic that "teamwork alone wins." As he opined, in the final analysis colleges and universities, as all institutions, survived only because of the good work they did and the goodwill they engendered. With Aber and other colleagues over the years in mind, Elrod extolled the qualities of the dedicated teacher, "seldom in print" and rarely recognized, who alone generated "the power of the machine." However, just as he worried about the animal and plant species threatened by the inexorable progress of civilization, "I fear present methods will cause him soon to disappear." In modern higher education, "Everyone tries to make money; few try to make men, outside of the family circle."[59]

And yet, as always, he exhorted and fiercely yearned for the younger generations to internalize the Aber and Ryman spirit for a creative resurgence of engagement, dedication to principle, and community life. In Elrod's view, the fervent desire and collaborative efforts of faculty and students inspired by a shared vision sufficed to shape a new reality. For himself, however, only misty memories of a golden past inspired the strength to endure the severe conditions and inexorable burdens of a dreary present. Magnifying the challenges threatening to overwhelm him, he no longer had access to the vanishing wilderness of the Biological Station at Flathead Lake or Glacier National Park. As he reminisced fervently in 1928 of the good times at the cherished station, "I missed those we had there, and those I love."[60]

V

A paralytic stroke in June 1934 required Elrod to use a cane or a wheelchair at all times, deprived him of the ability to speak or write coherently, and tragically ended his career as a naturalist-educator.[61] With some lingering hope for at least a partial recovery, the devastating impact of the stroke fully disabled him. His daughter Mary always insisted that overwork and stress caused the stroke and never changed her view, even after Elrod's death nearly two decades later, in 1953.[62] He left no personal record of life after 1934, with Emma and Mary attending to the life needs beyond his capacity. His daughter commented about the anguish caused by the loss of any opportunity to talk of the early years with her father.[63] A few photographs in the Elrod papers reveal little, except his reliance on the cane, perhaps to avoid the wheelchair, and an apparent desire to revisit familiar places such as Glacier National Park and the Biological Station at Flathead Lake.[64] Without doubt, the affliction wounded him, a man whose entire life had revolved around robust physical and intellectual activity. Life still had a few hard blows to deliver as Elrod endured the inactive silence of his infirmity for nearly twenty years.

Since retirement income depended primarily on the individual, prior to the Social Security Act of 1935, Elrod and his wife Emma relied heavily on the salary of daughter Mary, some accumulated but very modest savings, and any contribution the university made.[65] When she learned of her father's stroke, Mary returned immediately to Missoula to assist, secured a half-time appointment as secretary to the dean of women, and became acting dean of women and an assistant professor in 1935.[66] Years later, Vice President R. H. Jesse explained that President Clapp arranged Mary's employment, after the failure of attempts to retain Elrod on the payroll, as "a scheme to provide a pension for her father."[67] For a few months before his own untimely death, President Clapp kept Elrod's name on the list of active faculty members with the rank of professor emeritus at a reduced salary of $400 annually, hardly a sufficient amount of money to support the family. Even before Clapp's death, the board denied the reduced salary for Elrod, and Mary's salary as an employee became more critical.[68]

Partially as a result of lingering health problems and old age, Emma died in 1938, leaving her husband and daughter on their own.[69] That same year, Elrod received an honorary doctor of laws from the university in recognition of his distinguished years of service and accomplishments.[70] He attended the ceremony but did not speak. Also honored the same day, former Professor Frederick Scheuch, a member of the original faculty of the university, delivered an address exhorting students to engage actively in their education, a theme Elrod perhaps silently endorsed. The honorary degree conveyed its own satisfaction but did little to sustain the family. Elrod probably qualified for a retirement lump sum payment under the federal Social Security Act of 1935, but no evidence indicates that he claimed it.

In 1939, the university included the names of Elrod and three other emeritus faculty members on the list of faculty eligible for annuities as special cases under the state's Teachers Retirement System (TRS) opened to university faculty that year.[71] With university support, Mary applied unsuccessfully for the basic TRS annuity for her father in 1939, claiming the special case exemption and asserting that overwork and stress had caused the stroke.[72] When she applied again in 1940, President G. Finlay Simmons explained in his supporting letter that the university considered Elrod as the "one member of the faculty here most deserving of the benefits of the System." The TRS board agreed and accepted the special case argument to grant an annuity for Elrod.[73] A few years later and perhaps as an after thought, in 1947 the TRS board presented an elaborate retirement certificate to Elrod designed by Irvin "Shorty" Shope, the Montana artist and fine arts graduate of the university, commemorating Elrod's retirement in 1941 after thirty-seven years of distinguished service.[74] He actually retired in 1934 with thirty-seven years of service, and received the annuity in 1940 not 1941. Even discounting the errors, and while significantly different in many respects, the handsome certificate was not unlike the traditional gold watch, but the annuity did make a difference in his remaining years.

Mary became acting dean of women when the dean resigned, and she held the interim position until 1936, when President Simmons appointed her assistant dean of women and assistant professor, to serve as acting dean until the appointment of a permanent dean of women.[75] However, she performed poorly because of her attention to the needs of her father rather than to her official duties and responsibilities. She also had a prickly personality and a habit of having her way (perhaps due to her being an only child in a very close-knit family). Her role as the dean of women when in loco parentis prevailed on campus, especially for female students, imposed its own particular demands and frustrations.[76] President Simmons found himself immersed in other problems related to his presidency, and so Mary remained acting dean for ten years.[77]

In 1946, after familiarizing himself with the university, new President James A. McCain changed Mary's assignment to assistant director of the museum and northwest history collection at her existing salary.[78] Anticipating some resistance, he reported that, to his surprise, she loathed the position of dean. Vice President Jesse once again commented in a confidential memorandum to her file that the president and others knew prior to the transfer that Mary lacked competency as a museum assistant.[79] Jesse suspected as well that she took advantage of the position and claimed far more privileges and flexibility in work schedules and duties than reason or campus regulations allowed. As fitting as the role in the museum appeared for the daughter of the founder of the university's original museum, she failed in that position as in the earlier ones, undoubtedly in great part because of dedicated care for her father. To end the drain on the university, President Carl McFarland notified her that she had to retire on 31 December 1956 because of financial exigency, but then he delayed the date for four months to allow her to enhance her Social Security benefits. Two decades after Clapp arranged an appointment and salary for her as a tribute to her father, and four years after her father's death, she finally retired from university employment.[80] The university fulfilled President Clapp's commitment as the only way to honor one of its most distinguished faculty members so tragically stricken before his time.

VI

From 1934 until his death in 1953, Elrod enjoyed the care and attention of his daughter Mary, who handled his affairs and made life bearable for him after tragedy struck. With less than full attention to her university duties, she arranged picnics at Yellow Bay, automobile trips to Glacier National Park and other favorite places, and managed his correspondence. During that time, she also organized his papers and thousands of photographs for deposit in the K. Ross Toole Archives of The University of Montana, purchased by the Friends of the Library in the 1970s. Some of the materials and Elrod's cameras went to the Biological Station at Flathead Lake.[81] In the end, the old naturalist, with his daughter's assistance, bestowed a wonderful legacy on his adopted university and the Biological Station at Flathead Lake.

The Missoula chapter of the Audubon Society, founded early in the century by Elrod, proposed a birdbath on the campus to honor him.[82] The Campus Development Committee approved the birdbath despite university policy against such memorials. According to J. W. Severy, chairman of the committee and Elrod's successor as chairman of the biology department, the committee approved the exception because of Elrod's long association with the university and the failure of the earlier designation of the campus as the Elrod Bird Sanctuary. In 1940, Campus Engineer T. G. Swearingen persuaded the Audubon Society, the committee, and President Simmons to subsidize the construction of a sundial instead of a birdbath to avoid the problem of getting water to a birdbath.[83] The Elrod Sundial, consisting of a concrete pedestal with a copper dial and gnomon, retains an honored place today on the campus, adjacent to the Oval between Craig Hall (Mathematics), one of the oldest structures, and the Payne Family Native American Center, one of the newest. As happened throughout his long career, success for an Elrod project demanded persistence, sometimes spanning the history of the institution he helped to shape.

Park Superintendent J. W. Emmert sought conversations with Elrod and access to his Glacier National Park photographs in 1952 to assess changes in the glaciers over the years.[84] According to Mary, Elrod agreed, although conversation seemed unlikely in view of his infirmity. In any event, she urged haste because of Elrod's waning strength. She had taken the summer off to attend to his increasingly fragile health, and they had recently celebrated his eighty-ninth birthday with a picnic at the Biological Station. That same year, Mary delivered the keynote speech for the rededication of the Morton J. Elrod School in Kalispell, replacing a structure destroyed by fire.[85] On 8 February 1956, three years after his death, the university announced the renaming of South Hall as Elrod Hall, an honor of great significance to him, had he lived to enjoy it, because it occurred on Charter Day.[86] Most significant of all, in view of Elrod's obsession throughout his professional career in Montana, Biological Station Director Richard Solberg presided in 1967 at the dedication of the Morton J. Elrod Laboratory at Yellow Bay, constructed with support from a National Science Foundation grant. The naming of the Elrod Laboratory that replaced the one he had designed in 1912 marked yet another new beginning for the station he had conceived and built.[87]

Accolades and honors for the renowned naturalist-educator proliferated for years before and after his death in 1953.

Over the years, he had accepted invitations to become a contributing member of a host of professional societies and organizations, but always with the intention of active engagement.[88] When he thought it impossible or difficult to participate actively, he declined, as he did the invitation in 1916 to become a Fellow of the American Geographical Society.[89] By way of contrast, he accepted immediately when invited to represent the western states on the Ecological Society of America's Committee on the Preservation of Natural Conditions for Ecological Study, in all likelihood because he recognized a moral imperative.[90] Similarly, he typically complied with requests from organizations such as the Smithsonian Institution for biological specimens from or information about Montana.[91]

In spite of these honors, life delivered one last blow before his death. In response to a revised résumé his daughter sent at his request, Sydelle M. Haskell apologized for rejecting it for inclusion in *Who's Who in America* because space constraints disallowed résumés of scholars no longer of "current national reference interest."[92] However, Elrod perhaps found solace in Haskell's pledge to include his résumé in *Who's Who in the West,* the region he had chosen in 1896 for his life's work, to avoid becoming an itinerant professor.

Postscript

During his active years from 1897 to 1934, Morton John Elrod earned a rightful place as one of Montana's most distinguished naturalists and educators. His stature as the state's premier nature photographer remains unsurpassed today, in large measure because of his remarkable productivity and primacy on the ground. Curiously, he has received little scholarly attention despite his productive career of thirty-seven years as a naturalist, educator, scientist, and photographer, and despite the voluminous collection of his papers. Without question, his contributions to shared governance and academic freedom proved critical to strong traditions on the Missoula campus. Equally important, Elrod and his colleagues, scholars such as William Draper Harkins and H. G. Merriam, insisted upon intellectual engagement as the measure of a faculty member, even without adequate financial support. Their civic and scholarly examples paid huge institutional dividends as The University of Montana matured into a graduate research university in the years after the stroke ended Elrod's career.

Most biologists know of the Biological Station he founded in 1898 at Flathead Lake and fiercely protected over the years. Unfortunately, economic hard times and the lack of institutional support delayed the realization of his vision for the station. However, in the years after World War II, the station became a center of excellence within The University of Montana and has attained even higher scientific renown in the twenty-first century. Succeeding station directors have internalized, expanded, and fulfilled Elrod's vision of the station as a haven where kindred souls might study nature. In that regard, a station faculty member secured in

2014 the largest research grant ever awarded to the university. Ranked today as one of the top two freshwater stations in the world, Elrod's Biological Station at Flathead Lake continues to burnish its deserved reputation.

Not as many people know of Elrod's role in the establishment of the National Bison Range in the Ravalli hills, which remains one of the great scientific and tourist attractions in Montana. Fewer still know of his early involvement in the effort to establish Glacier National Park and his role in the development of the Park Naturalist Service, familiar to the millions of people from around the world who have visited the park. Not many scholars have consulted his collected papers or perused his extensive popular and scholarly work on the natural wonders and life forms of Montana, Flathead Lake, and Glacier National Park. In a long and interesting career, he contributed significantly to the knowledge of his adopted state and to the science of nature.

Finally, his attention to students—the quality of the education they received, and their preparation and academic engagement while at the university—helped to shape an enduring institutional focus on student achievement. In a characteristic campus-oriented and student-centered gesture in 1915, he implored the Missoula City Council to restore the lamp on Maurice Avenue at the point where the streetcar stopped, a beacon that signaled entry to campus.[1] Throughout his productive career, he exemplified the symbiosis encapsulated in the motto of The University of Montana, *Lux et Veritas*.[2] Because of his example and the dedication of generations of administrators, faculty, and students, founding president Oscar John Craig's prophetic epigram endures: "The University of Montana, it shall prosper."[3]

Appendix

Publications of The University of Montana Biological Station

Chronology of University Bulletins and University of Montana Studies

1900. Earl Douglass, "The Neocene Lake Beds of Western Montana."

1901. P. M. Silloway, "Summer Birds of Flathead Lake."

1902. M. J. Elrod, "A Biological Reconnaissance in the Vicinity of Flathead Lake."

1903. M. J. Elrod, "Lectures at Flathead Lake."

1903. P. M. Silloway, "Additional Notes to Summer Birds of Flathead Lake."

1904. M. J. Elrod, "Birds and Their Relation to Agriculture."

1904. M. J. Elrod, "The Resources of Montana and Their Development."

1904. M. J. Elrod, "The University of Montana Biological Station and Its Work."

1904. W. P. and C. W. Harris, "Lichens and Mosses of Montana."

1906. M. J. Elrod, "The Butterflies of Montana."

1906. J. A. Henshall, "A List of Fishes of Flathead Lake."

1908. M. J. Elrod, "Pictured Rocks."

1910. Marcus E. Jones, "Montana Botany Notes."

1920s. J. E. Kirkwood, "Botanical Science in Relation to Human Welfare."

1920s. J. E. Kirkwood, "The Comparative Embryology of the Cucurbitaceae."

Prior to 1924. J. E. Kirkwood, "Forest Distributions in the Northern Rocky Mountains."

Pamphlets and Books Produced at the Biological Station

A. W. L. Bray, *Protozoa of Flathead Lake and Vicinity,* 2,000 copies.
Lee R. Dice, *Mammal Associations of Flathead Lake,* 2,000 copies.
M. P. Dunkle, *Plant Communities of the New Road at Yellow Bay,* 2,000
    copies.
M. J. Elrod, *Flathead Lake Bird Reserve,* 2,000 copies.
M. J. Elrod, *Flathead Lake, Precipitation and Evaporation.*
M. J. Elrod, *Flathead Lake: Rise and Fall; Surrounding Region.*
M. J. Elrod, *The Food of the Fishes at Flathead Lake,* 2,000 copies.
M. J. Elrod, *Habits of the Richardson Squirrel.*
M. J. Elrod, *Observations of the Work of Barkbeetles of the Species*
    Dendractonas *in a Yellow Bay Pine Tree at the University*
    Biological Station.
M. J. Elrod, *Photography in the Mountains,* 2,000 copies.
M. J. Elrod, *Some Additional Lakes of Glacier National Park,* 2,000
    copies.
M. J. Elrod and G. B. Claycomb, *The* Entomastraca *of Flathead Lake*
    and the Relation to Fish Food.
M. J. Elrod and Francis Ross, *The Fishes of Flathead Lake.*
Marcus E. Jones, *The Flora of Western Montana,* 2,000 copies.
    Unpublished. 375 pages, 25 photographs.
Marcus E. Jones, *Notes on the Flora of Glacier National Park,* 2,000
    copies.
Gertrude P. Norton, *Shore Vegetation of Flathead Lake,* 2,000 copies.
P. M. Silloway, *Notes on the Birds of Glacier Park,* 2,000 copies.

Published Articles, Pamphlets, and Books

All titles by M. J. Elrod unless otherwise noted.

"Among the Kootenais," *Rocky Mountain Magazine* 3 (1901): 171–85.
    9 illustrations, 1 plate.
"The Beauties of the Mission Range," *Rocky Mountain Magazine* 2
    (1901): 623–31.
"Biological Tables," *Journal of Applied Microscopy* 2 (4) (no date).
"Collecting Shells in Montana," *The Nautilus* 15 (1902). 3 papers.

"The Flathead Buffalo Range: A Report to the American Bison Society," American Bison Society, *Annual Report* (1908): 15–49.

"Further Notes on the Use of Telephoto Lens," *Journal of Applied Microscopy* (1901): 1568–71. 5 figures.

H. W. Whitford, "Forests of the Flathead Valley," *Botanical Gazette* (1905). 23 illustrations, 1 map.

"The Lakes of Glacier National Park: Avalanche," *Transactions of the American Microscopical Society* (1910).

"The Lakes of Glacier National Park: Louise," *Transactions of the American Microscopical Society* (1912).

"Limnological Investigations at Flathead Lake," *Transactions of the American Microscopical Society* 12 (1899): 63–80. 9 plates.

M. J. Elrod & Maurice Ricker, "A New Hydra," *Transactions of the American Microscopical Society* (1901): 257–58.

"The Montana Bison Range," *Journal of Mammalogy,* 7:1 (February 1926): 45–48.

"Montana Shells," *Rocky Mountain Magazine* 2 (1901): 691–97. 4 plates.

"The Passing of the Pablo Buffalo Herd," *Shields' Magazine* 12:2 (February 1911): 35–41.

Some Lakes of Glacier Park (Washington, DC: Department of the Interior, 1912).

"The University of Montana Biological Station," *Journal of Applied Microscopy* 4 (5) (no date).

"The Value of Telephoto Lens," *Journal of Applied Microscopy* (1901): 1241–42. 2 figures.

Notes

Abbreviations

B	Box
CAD	Clyde Augustus Duniway: Papers Relating to the University of Montana, 1908–1912, Microfilm, MSS 735, K. Ross Toole Archives, Maureen and Mike Mansfield Library, University of Montana.
F	Folder or file
GBG	George Bird Grinnell Papers, Microfilm, Maureen and Mike Mansfield Library, University of Montana.
Human Resources	Human Resource Records, Record Group Number 30, K. Ross Toole Archives, Maureen and Mike Mansfield Library, University of Montana.
Inter-Mountain Educator	*Inter-Mountain Educator,* K. Ross Toole Archives, Maureen and Mike Mansfield Library, University of Montana.
JMD	Joseph M. Dixon Papers, K. Ross Toole Archives, Maureen and Mike Mansfield Library, University of Montana.
M	Microfilm
Missoulian	*Missoulian,* Maureen and Mike Mansfield Library, University of Montana, or Montana Historical Society, Digitized Newspapers, Helena, Montana.
MHS DN	Montana Historical Society, Digitized Newspapers, Helena, Montana, at mhs.mt.gov/research/collections/newspapers/dignews.
MJE	Morton J. Elrod Papers, Number 486, K. Ross Toole Archives, Maureen and Mike Mansfield Library, University of Montana.

MSS   Manuscripts
PO   Office of the President, Record Group Number 1,
                     K. Ross Toole Archives, Maureen and Mike
                     Mansfield Library, University of Montana.
R   Reel
RG   Record Group
S   Series
SBE   State Board of Education, Maureen and Mike
                     Mansfield Library, University of Montana.
UniPub   University Publications, President's Reports,
                     bound, K. Ross Toole Archives, Maureen and
                     Mike Mansfield Library, University of
                     Montana.
Weekly Missoulian   *Weekly Missoulian,* Maureen and Mike Mansfield
                     Library, University of Montana.

Introduction

1. For background, see John C. Beard, "Morton J. Elrod and The University of Montana Biological Station" (History 598: The Trans-Mississippi West, April 1968, University of Montana), Section 1, MJE, S1, B1, F1; Morton J. Elrod, "Many Changes in the City Are Told by Elrod," *Missoulian* (29 January 1922), 1, 5, MJE, S3, B7, F14; "A Great Strike. Motormen on All the Electric Street Cars Go Out," *Weekly Missoulian* (22 February 1893), Jan. 4, 1893–Dec. 26, 1894; and "Initial Run of Missoula Street Cars Proves Success and Pleases Many," *Missoulian* (12 May 1910), R39, an expansion of service.

2. Michael P. Malone and Richard B. Roeder, *Montana: A History of Two Centuries* (Seattle and London: University of Washington Press, 1988), ch. 5.

3. T. C. Spaulding, Dean of Forestry, "Forestry Education in Montana" (no date but early 1920s), RG1, PO, S15, B32, F "Forestry, School of, 1914–1934."

4. James Willard Hurst, *Law and the Conditions of Freedom in the Nineteenth-Century United States* (Madison: The University of Wisconsin Press, 1956), passim.

5. Malone and Roeder, *Montana,* ch. 8.

6. Morton J. Elrod, "George Bird Grinnell" (no date but probably 1928–29), handwritten and typescript (57 pages), for the quotation, MJE, S4, B16, F4–5; and Morton J. Elrod, handwritten notes on Grinnell's 18 articles in *Forest and Stream,* MJE, S4, B16, F6–7. On the Grinnell comment about Indian starvation, see Joseph Kinsey Howard, *Montana: High, Wide, and Handsome* (Reprint ed.; Lincoln and London: University of Nebraska Press, 1983), 155–56; and Gerald A. Diettert, *Grinnell's Glacier: George Bird Grinnell and Glacier National Park* (Missoula: Mountain Press Publishing Co., 1992), esp. 21. See also Morton J. Elrod, handwritten notes on and a copy of George Bird Grinnell, "The

Crown of the Continent," *Century Magazine* no. 76 (September 1901), 660–72, MJE, S4, B16, F9; "Glacier Centennial: George Bird Grinnell," *Malcolm's Round Table,* at http://knightofswords.wordpress.com/2010/03/28/glacier-centennial -george-bird-grinnell/; and "George Bird Grinnell," at http://en.wikipedia.org /wiki/George_Bird_Grinnell.

7. Dawes Act at http://en.wikipedia.org/wiki/Dawes_Act; Malone and Roeder, *Montana,* 270–71; and Burton Smith, "The Politics of Allotment: The Flathead Reservation as a Test Case," *Pacific Northwest Quarterly* 70, no. 3 (July 1979), 130–40.

8. Malone and Roeder, *Montana,* 124–28.

9. Ibid., ch. 10–12 and 15; also see "Montana," at http://en.wikipedia.org /wiki/Montana.

10. Thomas J. Dimsdale, *Vigilantes of Montana or Popular Justice in the Rocky Mountains,* 3rd ed. (Helena, MT: State Publishing Co., 1915), passim.

11. "Montana," at http://en.wikipedia.org/wiki/Montana; also Howard, *Montana: High, Wide, and Handsome,* ch. 16–21.

12. Malone and Roeder, *Montana,* 217–19, and ch. 9 and 12; and Arnon Gutfeld, *Montana's Agony: Years of War and Hysteria, 1917–1921* (Gainesville, FL: University Presses of Florida, 1979), ch. 9 and 12.

13. Robert H. Wiebe, *Search for Order: 1977–1920* (New York: Hill and Wang, 1967), passim. On the concentration of wealth, see Thomas Piketty, *Capital in the Twenty-First Century* (Cambridge and London: Belknap Press of Harvard University Press, 2014), 347–53 passim.

14. Malone and Roeder, *Montana,* ch. 9–12; Wiebe, *Search for Order,* ch. 5–9; and Gutfeld, *Montana's Agony,* esp. introduction. Also Thomas C. Leonard, *Illiberal Reformers: Race, Eugenics, and American Economics in the Progressive Era* (Princeton, NJ: Princeton University Press, 2016), parts 1–2.

15. Michael P. Malone, *Battle for Butte: Mining and Politics on the Northern Frontier, 1864–1906* (Seattle: University of Washington Press, 1981), passim.

16. Malone and Roeder, *Montana,* 158, 280–82. For example, Sheila MacDonald Stearns, "The Arthur Fisher Case" (Master's thesis, University of Montana, Missoula, 1969), iv, passim, esp. 14–25. For the rise of the Anaconda Company, see Gutfeld, *Montana's Agony,* ch. 1.

17. Gutfeld, *Montana's Agony,* ch. 5 and 12.

18. Malone and Roeder, *Montana,* 276–77; and "Thanks, Gentlemen," *Weekly Missoulian* (3 February 1893), R Jan. 4, 1893–Dec. 26, 1894.

19. As explained years later by Chancellor George A. Selke, "The Needs of The University of Montana," 8 January 1947, RG1, PO, S19, B52, F "Chancellor, Selke, George A. 1946–48." See also Michael P. Malone, "The Montana University System: The First Half Century," *Montana: The Magazine of Western History* 44, no. 2 (Spring 1994), 60–64; George M. Dennison, "Higher Education in Montana, 1950–1993," *Montana: The Magazine of Western History* 44, no. 2 (Spring 1994), 65–72.

20. Edward L. Pike, "Needed: Consolidation of Montana's University System Into Three Rather Than Six Units," *Lewistown Daily News* (3 August 1958), RG1, PO, S19, B52, F "1958 Bonding Issue."

21. See various issues of the *Weekly Missoulian* (January–February 1893), R Jan. 4 1893–26 Dec. 1894, esp. "A Helena Scheme," "Scotch'd the Job," "The University Club," "The Merry War" (all 11 January 1893), and "The University" (25 January 1893); also James M. Hamilton, "Early History of the University of Montana," Charter Day Address, 17 February 1925, esp. 9, 13–14, and 18; and "First Movement For The Establishment of The University of Montana As Told By E. E. Hershey to Paul C. Phillips" (no date, but probably 1930s), both RG1, PO, S2, B25, F "ARCHIVES, History." See also William Giltner, "Montana State University," 15 February, 1939, esp. 1–5; Norma Beatty, "Montana University's Birth Saw Big Debate, Lobbying," *Kaimin* (8 April 1957), 3, 5 (excerpt); and Fred G. Scheuch, "University's Start Was Very Modest," *Missoulian, Souvenir Edition* (20 July 1922), passim, esp. 10–11, all RG1, PO, S15, B26, F "ARCHIVES: Informational."

22. A. L. Stone, "An Address Delivered On The University of Montana Charter Day," 1914, RG1, PO, S4, B167, F "Charter Day—Addresses, 1912, 1914, 1918, 1919, 1920, 1921." See also "In the Senate," *Weekly Missoulian* (5 January 1893), for the comment that the consolidation "proposition may be said to have already failed"; "Rah for the Varsity," *Weekly Missoulian* (31 January 1893); "Thanks, Gentlemen," *Weekly Missoulian* (3 February 1893); and "Same Old Thing," *Weekly Missoulian* (17 February 1893), all R Jan. 4, 1893–Dec. 26, 1894. With vote trading in the effort to elect a Montana senator, Matts succeeded in the argument for the university at Missoula and that defeated the consolidation effort.

23. H. G. Merriam, *University of Montana: A History* (Missoula: University of Montana Press, 1970), "Appendix," the enabling act, dated 17 February 1893, 183–85. The University Club members drafted the act; see *Weekly Missoulian* (14 January 1893), R Jan. 4, 1893–Dec. 26, 1894.

24. Mary Brennan Clapp, "Narrative of Montana State University: 1893–1935," 1958, typescript (copies in author's possession and in the Toole Archives), ch. 2: 7, 11–14.

25. Oscar J. Craig, "President's Report," 1896, 5, UniPub, S2, B1895–1912; Elrod, "Many Changes," *Missoulian* (29 January 1922), 1, 5, MJE, S3, B7, F14; Merriam, *History*, 38; J. M. Hamilton to Morton J. Elrod, 20 December 1929, MJE, S2, B3, F2; and Morton J. Elrod, "Education," Address to the Quarter Century Club (no date, but probably mid-1920s), MJE, S8, B36, F9.

26. "Morton J. Elrod Biography," http://bio19c.com/-biography153_morton_john_elrod_(1863–1953); also "Prof. M. J. Elrod," *Wesleyan Echo* (1896) at http://collections.carli.illinois.edu/u?/iwu_argus,36438; and "Guide to the Morton J. Elrod Papers, 1885–1959," for a brief biography and bibliography, at http://archiveswest.orbiscascade.org/ark:/80444/xv45831.

27. Mary Elrod Ferguson to J. E. Thornton, 9 March 1954, MJE, S1 B1, F5. The Elrod papers contain a lengthy genealogy, MJE, S1, B1.

28. "Elrod, Morton J.," at http://collections.carli.illinois.edu/u?/iwu_histph ,2464. Also *Announcement of the Graduate and Non-Resident Department of the Illinois Wesleyan University* (Bloomington: Illinois Wesleyan University, 1904), esp. 36–39; "Biology" under "Degrees Conferred," *Quarterly Bulletin of the Illinois Wesleyan University* (Bloomington: Illinois Wesleyan University, 1905), 104, "Elrod, Morton J., Missoula, MT. (A.B., Drake University [error]), Biology;" and see Henry Christopher Allan, Abstract of "History of the Non-Resident Degree Program at Illinois Wesleyan University, 1873–1910: A study of a Pioneer External Degree Program in the United States" (PhD diss., University of Chicago, 1984, copy in author's possession), passim, the first external degree program in the United States. See also Beard, "Morton J. Elrod," Section 1, 1–2; Charles C. Adams, "Memorandum on M. J. Elrod's Start as a Teacher," 20 October 1892 (stamped 30 June 1949), MJE, S1, B1, F1; and Alice Elrod, "MJ Elrod Items" (no date but 1950s), MJE, S1, B1, F1–1.

29. *The Wesleyan . . . of The Illinois Wesleyan University,* I (Bloomington, IL: Students, 1895), 37, 45–47, MJE, S1, B1, F7; and "Prof. M. J. Elrod," *Wesleyan Echo* (1896), at http://collections.carli.illinois.edu/u?/iwu_argus,36438; also "Annual Meeting of the Joint Board of Trustees and Visitors," Illinois Wesleyan University, 10 June 1889, copy in author's possession.

30. M. J. Elrod to F. C. Scheuch, 6 May 1916, RG1, PO, S15, B26, F "Museum, 1915–1944." And see "Prof. M. J. Elrod," *Wesleyan Echo* (1896) at http://collections.carli.illinois.edu/u?/iwu_argus,36438.

31. "Elrod Family Records," MJE, S1, B1, F2; and Mary Elrod Ferguson, typescript summary of Elrod's Life (no date but 1950s), MJE, S1, B1, F1.

32. Mrs. C. E. Green to Mrs. Elrod, 3 September 1898, MJE, S1, B1, F8.

33. Morton J. Elrod to Emma and Mary, 24 August 1922, MJE, S2, B22, F7.

34. Morton J. Elrod to Emma, 16 July 1924, Emma's salary of $100, MJE, S2, B2, F9.

35. See, for example, Emma H. Elrod, "Flathead Indian Reservation," in *Woman's Souvenir of Missoula Montana,* edited by Elizabeth L. Mills (Missoula: Ladies of the Christian Church, 1910), at 18–19, Special Collections, K. Ross Toole Archives, Maureen and Mike Mansfield Library, The University of Montana.

36. *Wesleyan,* 45–47, 94–98, MJE, S1, B1, F7. Also Morton John Elrod, "Among the Rockies," *Wesleyan Argus* 1, no. 2 (28 September 1894), 5–7; no. 5 (12 November 1894), 5–8; no. 10 (28 January 1895), 5–9; and no. 16 (28 April 1895), 6–9; all MJE, S8, B37, F19.

37. Beard, "Morton J. Elrod," Section 1, 1, and Section 2, 4–6; Charles (Adams) to Morton J. Elrod, 17 July 1893, MJE, S2, B1, F8; and *Wesleyan,* 45–47, MJE, S1, B1, F7. Also Morton J. Elrod, "A Camp at Mt. Lolo," *Outdoor World* (1899), 65–68; and "The Ascent of Mount Lolo," *Illinois Wesleyan Magazine*

(1893), 322–28, both MJE, S3, B7, F9 and F10, respectively; also excerpts from University of Montana President's Report, 30 May 1896, MJE, S5, B34, F10; and Clapp, "Narrative," ch. 2, 29–30.

38. Elrod, "Many Changes," *Missoulian* (29 January 1922), 1, 5, MJE, S3, B7, F14; Beard, "Morton J. Elrod," Section 1, 1; Clapp, "Narrative," ch. 2, 29–30, 33–34; and Merriam, *History,* 6.

39. "Resolution of Joint Board of Trustees and Visitors," Illinois Wesleyan University, 15 December 1896, MJE, S2, B1, F8.

40. Morton J. Elrod, "Evolution and Religion Do Not Conflict, Say Teachers," *Missoulian* (31 May 1925), 1, 10, MJE, S6, B34, F17.

41. Elrod, "Many Changes," 1, 5, MJE, S3, B7, F14.

42. Morton J. Elrod, "Air Currents in Montana," 1904, MJE, S3, B7, F1.

43. Elrod, "Many Changes," 1, 5, MJE, S3, B7, F14. Also Morton J. Elrod, "Geological Description of Region" (no date, but early 1900s), MJE, S3, B7, F2; Morton J. Elrod, "Frye's Geography Reviewed," *Missoulian* (10 June 1912), MJE, S3, B7, F5; Morton J. Elrod, "Life Changes in Western Montana," paper for the Cosmos Club (a University-Missoula town-gown club meeting monthly except during summers to hear a scholarly paper or talk delivered by one of the members), 8 January 1911, MJE, S3, B7, F8; Morton J. Elrod, "Missoula County," January 1923, MJE, S3, B7, F13; Morton J. Elrod, "The Beauties of the Mission Range" (1901), offprint of article in *Rocky Mountain Magazine* 2, no. 2 (April 1901), 623–31, MJE, S3, B7, F12.

44. Clapp, "Narrative," ch. 2, 7–8.

45. Elrod, "Many Changes," 1, 5 MJE, S3, B7, F14.

46. Beard, "Morton J. Elrod," Section 2, 1–5; Merriam, *History,* 13; and Clapp, "Narrative," ch. 2, 54–57.

47. Beard, "Morton J. Elrod," Section 2, 5; and Merriam, *History,* 6.

48. For enrollments, faculty, and growth of the University, see Giltner, "Montana State University," 15 February 1939; and Scheuch, "University's Start Was Very Modest," *Missoulian, Souvenir Edition* (20 July 1922), 9–12, both RG1, PO, B26, F "ARCHIVES: Informational"; Beard, "Morton J. Elrod," Section 2, 4–5; also Elrod, "Many Changes," 1, 5, MJE, S3, B7, F14; Morton J. Elrod, "First [1897] Football Team, University of Montana" (no date, but probably in the mid- to late 1920s), MJE, S5, B27, F11; and Clapp, "Narrative," ch. 2, 34.

49. Clapp, "Narrative," ch. 2; and Beard, "Morton J. Elrod," Sections 1–2. See as well Arthur L. Stone, *Following Old Trails* (Missoula: Morton John Elrod, 1913), passim (available online at http://archive.org/details/followingold trai00ston).

50. "Naturalism (philosophy)," at http://en.wikipedia.org/wiki/Naturalism _(philosophy), especially "methodological naturalism," on the means of discovering and verifying knowledge of nature, the only real knowledge. For the origins of this approach, David Wootton, *Invention of Science: A New History of the Scientific Revolution* (New York: Harper Collins, 2015), ch. 1–2. The following

discussion adopts the usage of Ullica Segerstralle, *Nature's Oracle: The Life and Work of W. D. Hamilton* (Oxford, UK: Oxford University Press, 2013), passim, esp. 2–3, to invoke the naturalist's fascination for, love of, and "profound empathy with all living things." This passionate "affinity with the natural world" facilitated and pervaded Elrod's research and teaching. However, methodological naturalists differ fundamentally from "metaphysical" or "ontological" naturalists because of their insistent reference of all knowledge to the laws of nature. See also Noam Chomsky's discussion of the application of methodological naturalism to the study of language and mind in *New Horizons in the Study of Language and Mind* (Cambridge: Cambridge University Press, 2000), passim.

51. Morton John Elrod, *Butterflies of Montana: With Keys for Determination of Species* (Missoula: University of Montana, 1906), 113.

52. Morton J. Elrod to Anna C. Furst, Big Sandy (no month, but July or August 1916), MJE, S2, B2, F2, emphasis supplied.

53. For the following discussion and the extended quotation, see Morton J. Elrod, "Mesology," or "The theory of organic evolution," Cosmos Club paper (undated, but 1908–1910), MJE, S8, B37, F1; and Morton J. Elrod, "Microbiology," Cosmos Club paper (February 1913), MJE, S8, B37, F9. See also "Alfred Russell Wallace," at http://en.wikipedia.org/wiki/Alfred_Russel_Wallace; and "Charles Darwin," at http://en.wikipedia.org/wiki/Charles_Darwin.

54. Morton J. Elrod, "The Value of Nature Study," Cosmos Club paper (no date but 1900–1910); and "On the Use of Reason," notes for a Cosmos Club paper (no date but 1900–1910), MJE, S8, B37, F12 and F18, respectively. Also see Chomsky, *New Horizons in the Study of Language and Mind,* passim.

55. "Ellsworth Huntington," at http://en.wikipedia.org/wiki/Ellsworth _Huntington. Elrod cited *Pulse of Asia.*

56. Beard, "Morton J. Elrod," Section 4, 5–8.

57. Morton J. Elrod, "The Three Seasons" (printed, no date but 1923), MJE, S8, B37, F14–15.

58. Morton J. Elrod, "Reason and Faith" (no date but 1920s); "Blazing the Trail" (no date but 1900–1910); "Old Places" (no date but 1900–1910), MJE, S8, B36, F3, F13, and F17, respectively; and "The Field Trip" (no date but 1910–15), MJE, S8, B37, F5.

59. Morton J. Elrod, "Accidents to birds" (August 1900), handwritten notes, MJE, S3, B9, F6.

60. Morton J. Elrod, *A Biological Reconnaissance in the Vicinity of Flathead Lake* (Reprint ed., Cornell University Library, Digital Collections, OH 105.M9E 48, no date), originally, *Bulletin University of Montana* (No. 10, Biological Series No. 3, 1902), 103–104, and 170–76, on "Montana Shells," with a bibliography, covering four research seasons; and Morton J. Elrod, "Effects of Altitude on Snails of the Species Pyramidula strigosa Gould," Section F, American Association for the Advancement of Science, Pittsburgh Meeting, "Abstract" in *Science* (29 August 1902), 349.

61. Elrod, *Biological Reconnaissance,* 155; and for the report and rejection of *hydra corala* for Elrod and Ricker, at http://www.marinespecies.org/aphia .php?p=taxdetails&id=564218.

62. "Prof. M. J. Elrod," *Wesleyan Echo* (1896), at http://collections.carli .illinois.edu/u?/iwu_argus,36438; also "Annual Meeting of the Joint Board of Trustees and Visitors," Illinois Wesleyan University, 10 June 1889, copy in author's possession; and Morton J. Elrod, "Department of Biology," in Oscar J. Craig, President's Report, 1903, 40–45, UniPub, S2, B1895–1912.

63. Morton J. Elrod, "The Heritage of Youth," 1928, Presidential Address, MJE, S3, B5, F16; Morton J. Elrod, "Minimum Essentials for High School Zoology," *Inter-Mountain Educator* 16, no. 7 (March 1921), 303–306; Morton J. Elrod, "Montana Teachers' Pension Law," *Inter-Mountain Educator* 10, no. 7 (March 1915), 24–25; and Elrod to Furst, Big Sandy (no month, but July or August 1916), MJE, S2, B2, F2.

64. See the discussion below of his service on AAUP subcommittees, and "State University Chapter," AAUP, 6 February 1933, MJE, S1, B4, F3.

65. See, for example, Elrod, "Effects of Altitude on Snails of the Species Pyramidula strigosa Gould," Section F, American Association for the Advancement of Science, Pittsburgh Meeting, "Abstract" in *Science* (29 August 1902), 349.

66. See esp. MJE, SI, B4, F3 and F7–8; and B13, F7; and various letters to Emma written during these trips scattered throughout the Elrod papers.

67. "Pyramidula Elrodi," at http://biostor.org/reference/130217; and H. A. Pilsby, The Academy of Natural Sciences of Philadelphia, to Morton J. Elrod, 19 and 30 July 1900, announcing "*Pyranidula Elrodi,*" MJE, S2, B1, F9; and "H. A. Pilsby," at https://books.google.com/books?id=ggY2AQAAMAAJ&lpg =PA173&dq=H.%20A.%20Pilsby&pg=PA173#v=onepage&q=H.%20A.%20 Pilsby&f=false.

68. See Clapp, "Narrative," ch. 2, 53; and "File: Procamelus elrodi molars1 jpg," at https://commons.wikimedia.org/wiki/File:Procamelus_elrodi_molars1 .jpg.

69. Elrod, "Geological Description of Region" (early 1900s), MJE, S3, B7, F2.

70. Ibid.; "Life Changes in Western Montana," Cosmos Club Paper (8 January 1911), MJE, S3, B7, F8.

71. "J. Harlan Bretz," at http://en.wikipedia.org/wiki/J_Harlen_Bretz; and David D. Alt, *Glacial Lake Missoula: And Its Humongous Floods* (Missoula: Mountain Press Publishing Co., 2001), passim.

72. For the following quotations, see Morton J. Elrod, *Vacation in Montana* (Bloomington, IL: The University Press, 1899), passim, MJE, S3, B7, F21. See also "Flathead Indian Reservation, at http://en.wikipedia.org/wiki /Flathead_Indian_Reservation.

73. The Dawes Act required legislation authorizing survey, allotment to Indians, and opening specific land to white settlement; see Jules Alexander Karlin, *Joseph M. Dixon of Montana* (Missoula: University of Montana, 1974), esp. vol. 1, 54–61, and vol. 2, 242; and Smith, "The Politics of Allotment," *Pacific Northwest Quarterly*, 130–40.

74. Morton J. Elrod, "List of Stories Sent to McClure Newspaper Syndicate, 246 West 59th Street, New York, N.Y." (no date but 1920s), MJE, S3, B5, F12. For a multipage listing of hundreds of stories, see MJE, SI, B8, F1–17. Also W. B. Parsons to Morton J. Elrod, 15 June 1916, MJE, S2, B2, F2. See as well Morton J. Elrod, "How the Blackfeet Indians Cook Eggs" (no date but 1910–15), MJE, S3, B6, F5; also Morton J. Elrod, several newspaper pieces, articles from Grinnell's *Forest and Stream,* and handwritten essays about Indians, particularly the Blackfeet (various dates), MJE, S3, B6, F2–12; and Morton J. Elrod, "Four Stories about the Flathead" (no date, but 1900–1908), MJE, S3, B5, F8.

75. Elrod, "Second Story," in Elrod, "Four Stories About the Flathead" (1900–1908), MJE, S3, B5, F8.

76. Morton J. Elrod, untitled request for museum consideration (undated but 1915 or 1916), MJE, S5, B28, F6.

77. Morton J. Elrod, "Where the Red Man Wrote His Story On the Rocks And Left The Record Of Great Salish Triumphs," *Missoulian* (1 March 1908), MJE, S3, B5, F10.

78. Robert G. Raymer to Morton J. Elrod, 7 August 1928, MJE, S2, B3, F1.

79. For Grinnell's influence, see Elrod, handwritten notes for a biography of Grinnell, a fifty-seven page speech on Grinnell, and a copy of Grinnell, "The Crown of the Continent," *Century Magazine* (1901), 660–72, MJE, S4, B16, F9.

80. Emma Elrod, "Flathead Indian Reservation," 19.

81. Morton J. Elrod, "Department of Biology" (undated, but 1916), MJE, S5, B23, F6.

82. Morton J. Elrod, "Conservation" (no date, but after 1924, probably 1925), MJE, S4, B15, F9; "The Passing of the Pablo Buffalo Herd," *Shields' Magazine* 12, no. 2 (February 1911), 35–41; "The Montana Bison Range," *Journal of Mammalogy* 7, no. 1 (February 1926), 45–48, both MJE, S3, B4, F12; and handwritten draft of report to the American Bison Society and other related fragments (undated but after 1910), MJE, S3, B4, F15.

83. Morton J. Elrod, "Brief History of the Montana Biological Station" (undated, but 1920), MJE, S5, B29, F5. See also Elrod, *Biological Reconnaissance,* 162–68, on the Swan Range.

84. "The Course of Empire" by Thomas Cole, at http://en.wikipedia.org/wiki/The_Course_of_Empire; and Leo Marx, *Machine in the Garden: Technology and the Pastoral Ideal in America* (London and New York: Oxford University Press, 1964), passim. Also see Morton J. Elrod, "The Proposed Glacier National Park," with "Photographs by the Author" (no date, but written prior

to passage of the congressional act establishing the Park in May 1910, probably early 1910, as indicated by a line through "Proposed" on the typed draft and in a folder marked 1907, with internal references to 1906 and 1909 excursions into the park region), MJE, S4, B13, F15, fifty-eight handwritten pages offering an eloquent word picture of the proposed park.

85. Victor E. Shelford, Ecological Society of America, to Morton J. Elrod, 7 July 1917, handwritten note, "ans. July 20," MJE, S2, B2, F3.

86. For the ambivalence, shared by George Bird Grinnell, see Jesse DeSanto, "Foreword," in Diettert, *Grinnell's Glacier,* xi–xii; and Andrew C. Harper, "Conceiving Nature: The Creation of Montana's Glacier National Park," *Montana: The Magazine of Western History,* 60, no. 2 (Summer 2010), 3–24, 91–94; shared as well by Charles M. Russell and many others, as see comments by Peter Hassrick and Brain Dippie, "Montana's Charlie Russell: 21st Century Perspectives on the Cowboy Artist," June 2015, 18–20, available online from the Montana Historical Society "Montana Memory Project"; for an analysis of the roots of the ambivalence, see Allan Bloom, *Closing of the American Mind* (New York: Simon and Schuster, 1967), esp. "Two Revolutions and Two States of Nature," 170–72.

87. Monthly deposits averaging about $250 (occasionally larger) in the 1920s; see Morton J. Elrod, Bank Transaction Records, 1920–21, MJE, S7, B35, F17.

88. Beard, "Morton J. Elrod," Section1, 1, and Section 2, 4–6.

89. William T. Hornaday, "Report of the President on the Founding of the Montana National Bison Herd," American Bison Society, *Second Annual Report of the American Bison Society, 1908–1909* (n.p.: American Bison Society, 1909), 1–19; and Elrod, "The Passing of the Pablo Buffalo Herd," *Shields Magazine* (February 1911), 35–41; also Elrod, "The Montana Bison Range," *Journal of Mammalogy* (February 1926), 45–48.

90. Morton J. Elrod, "Indian Pictures" (no date, but 1900–1910), MJE, S7, B37, F29; and orders to Dresden, Germany, at a cost allowing profit at fifty cents each, MJE, S7, B35, F10.

91. Morton J. Elrod to Emma, 29 July 1910, MJE, S2, B1, F10.

92. Morton J. Elrod to Emma, 21 July 1903, MJE, S2, B1, F9.

93. Morton J. Elrod, *Elrod's Guide and Book of Information of Glacier National Park* (2nd ed., revised and enlarged; Missoula: Morton J. Elrod, 1930), passim.

94. Morton J. Elrod, *Views of the Mission Mountains . . . Flathead Lake and Valley, Montana* (Missoula: Morton J. Elrod, 1908), page of text and hand-tied cover. See Albertype Book Co., NY, to Morton J. Elrod, 19 April 1913, printing costs, MJE, S7, B35, F10. Also Morton J. Elrod to Emma, 29 July 1910, MJE, S2, B1, F10.

95. Elrod, "Beauties of the Mission Range" (1901), 623–31, MJE, S3, B7, F12. Also Stone, *Following Old Trails,* 167; and Sam Johns, "A Journey to the

Flathead," September 1889, in Rick and Susie Graetz, "The Flathead Lake Story," e-mail communication, 24 September 2013, in author's possession.

96. Morton J. Elrod, handwritten notes in small notebook (no date, but late 1920s), seeds to people in various states totaling about $140.00, MJE, S7, B35, F17. Also Morton J. Elrod to J. Milne, Esq., Irvine House, Canonbie, Scotland, 4 July 1912; Morton J. Elrod to German dealer, 24 December 1912; and J. Robertson, England, to Morton J, Elrod (no date but 1912), all MJE, S7, B35, F14; and Morton J. Elrod to Clyde A. Duniway, 23 August 1908, tamarack seed for the Royal Botanical Garden, Kew, England, CAD, MSS 735, R4.

97. Investment Records (no dates but 1915–20), MJE, S7, B35, F1–2 and 12.

98. Morton J. Elrod, "The Inter-Mountain Educator," *Montana Education* 1, no. 3 (November 1924), 15–16; also Morton J. Elrod to Emma, 16 July 1924, Emma's salary of $100, and 28 July 1924, both MJE, S2, B2, F9; and Morton J. Elrod to Ira B. Fee, President, Montana Education Association, 31 December 1925, MJE, S2, B2, F10. See also Morton J. Elrod to William Ferguson, 15 August 1917; and Morton J. Elrod to Chancellor E. C. Elliott, 17 August 1917, both MJE, S2, B2, F3.

99. Stone, *Following Old Trails,* passim; Morton J. Elrod, various materials (no date but 1908 to 1916), MJE, S7, B35, F14; and Morton J. Elrod, "Following Old Trails," *Inter-Mountain Educator* 9, no. 6 (February 1914), 25–26; and A. L. Stone, Dean, to John Dexter, Referendum Campaign, 12 August 1930, RG1, PO, S15, B35, F "School of Journalism, 1911–1942."

100. Morton J. Elrod, F "Following Old Trails, 1913–1916," sales, MJE, S7, B35, F17.

101. Harold C. Urey to Morton J. Elrod, 21 November 1934, MJE, S2, B3, F8; and 6 July, 20 July, and 30 July 1917, all MJE, S2, B2, F3. On Urey, see "Harold Urey," http://en.wikipedia.org/wiki//HaroldUrey.

Chapter 1

1. "Morton J. Elrod Biography," http://bio19c.com/-biography153_morton _john_elrod_(1863–1953).

2. Founded in 1850, Illinois Wesleyan University no longer awards doctoral degrees, "Illinois Wesleyan University," at https://www.iwu.edu/aboutiwu /history.html. On the graduate "external" degree, see Allan, abstract of "History of the Non-Resident Degree Program at Illinois Wesleyan University, 1873–1910" (1984), copy in author's possession.

3. Walter Isaacson, *Einstein: His Life and Universe* (New York: Simon and Schuster, 2007), 68–72, 101–103.

4. Elrod, *Butterflies of Montana,* passim, with the assistance of an illustrator.

5. A. L. Strand, President, Montana State College, "Foreword," in *Bibliography of Graduate Theses, University of Montana* (Works Progress Administration,

Service Division, Historical Records Survey Project, Bozeman, MT, January 1942), at iii, erroneously claiming Montana State College awarded the first graduate degree, RG1, PO, S4, B63, F "Degrees at all University Units (Master's)"; also Morton J. Elrod, "Department of Biology," 1901–1902, MJE, S5, B23, F6; Morton J. Elrod, "Department of Biology," in Oscar J. Craig, President's Report, 1902, 44–46, UniPub, S2, B1895–1912; and for Douglass's thesis, "Numbered Publications," clipped file, March 1934, RG1, PO, S19, B46, F "Publications, Procedures for Distribution to State Board."

6. Richard E. Rice and George B. Kauffman, "William Draper Harkins: An Early Environmental Chemist in Montana (1900–1912)," *Bulletin for the History of Chemistry* 20 (1997), 60–67, at http://www.scs.illinois.edu/~mainzv /HIST/bulletin_open_access/num20/num20%20p60–67.pdf.

7. "Papers on smelter smoke," judge's decision and three scholarly papers, at http://www.23.us.archive.org/stream/cu31924051103566/cu31924051103566 .djvu.text.

8. Morton J. Elrod to Clyde A. Duniway, 6 July 1908, CAD, MSS 735, R4; and Morton J. Elrod to the McClure Newspaper Syndicate, New York, 24 April 1926, MJE, S3, B8, F1.

9. Oscar J. Craig, President's Report, 1897, 9, UniPub, S2, Box 1895–1912.

10. James M. Hamilton, "Early History of the University of Montana," 17 February 1925, esp. 9, 13–14, and 18, RG1, PO, S2, B25, F "ARCHIVES, History." Also Giltner, "Montana State University," 15 February, 1939, esp. 1–5; and Scheuch, "University's Start Was Very Modest," *Missoulian, Souvenir Edition* (20 July 1922), esp. 10–11, both RG1, PO, S15, B26, F "ARCHIVES: Informational."

11. J. M. Hamilton, J. W. Kleck, Oscar J. Craig, and D. W. Sanders, State Board of Education Special Committee, to State Board of Education, 28 November 1897, RG1, PO, S4, B168, F "Letters Concerning Vocations. . . ." In accord with his principles, Elrod's staff work included requirements for laboratory work in science.

12. Craig, President's Report, 1903, 13–15, UniPub, S2, B1895–1912. Also SBE, Minutes, Regular Meeting, 6 June 1904, 504–508, M418, R2; and H. A. Hollister, high school visitor, to President Oscar J. Craig, 14 September 1905, RG1, PO, S4, B168, F "Correspondence and Statistics Relating to Accredited High Schools—1905."

13. Elrod, "Many Changes," *Missoulian* (29 January 1922), 1, 5, MJE, S3, B7, F14.

14. Morton J. Elrod, "The American University" (no date but about 1916), MJE, S3, B4, F9.

15. Morton J. Elrod, "The Heritage of Youth," 1928, MJE, S3, B5, F16. The issue remains contentious today.

16. Oscar J. Craig, President's Report, 1905, 22; 1906, 11–13; and 1907, 16–23, UniPub, S2, B1895–1912.

17. Craig, President's Report, 1906, 11–13; and 1907, 16–24, UniPub, S2, B1895–1912. Also "University of Montana. Course of Study for Accredited High Schools" (no date, but 1906 or 1907), RG1, PO, S4, B168, F "Letters Concerning Vocations"; and Hollister to Craig, 14 September 1905, both RG1, PO S4, B168, F "Correspondence and Statistics Relating to Accredited High Schools—1905."

18. Craig, President's Report, 1907, 24, UniPub, S2, B1895–1912.

19. A. J. Stone, "An Injustice," *Missoulian* (2 December 1910), 4, MHS DN, for the Governor's veto of a similar bill. The University finally prevailed in 1911, "Some Good Bills Passed By Montana Legislature," *Missoulian* (3 March 1911), 1, 7, MHS, DN.

20. Morton J. Elrod to Emma, 15 July 1905, MJE, S2, B1, F9.

21. See letters to Bausch and Lomb, 1901–1907, RG1, PO, S4, B167, F "Correspondence—1901–1908."

22. Judith A. Rile, "The Changing Role of the President in Higher Education," 2001, at http://www.newfoundations.com/OrgTheory/Rile721.html. Also Oscar J. Craig to Prof. M. J. Elrod, 2 November 1901, RG1, PO, S4, B168, "F "Correspondence Received—Pres. O. J. Craig, about 1897–1907 . . . (A-R)"; and Neil W. Hamilton, *Academic Ethics: Problems and Materials on Professional Conduct and Shared Governance* (Washington, DC: American Council on Education and Praeger Publishers, 2002), passim.

23. William A. Aber to Clyde A. Duniway, 13 July 1908, CAD, MSS 735, R4; and Clapp, "Narrative," ch. 3, 6–8.

24. Clapp, "Narrative," ch. 3, 19–22. Also F. C. Scheuch, Acting President, "Data Concerning the State University," 1916, no pagination, RG1, PO, S15, B26, F "ARCHIVES: Informational"; and on the increasing duplication, see "At the Montana State University: This Week Marks An Important Epoch in the History of the Institution Across the River," *Missoulian* (30 August 1908), 3, R32.

25. Morton J. Elrod, "Department of Biology Report," 1903, MJE, S5, B23, F7.

26. Elrod, "Department of Biology," 1901–1902, MJE, S5, B23, F6; and Elrod, "Department of Biology," in Craig, President's Report, 1902, 44–46, UniPub, S2, B1895–1912.

27. Elrod, "Department of Biology," 1901–1902, MJE, S5, B23, F6; and Elrod, "Biological Station," in Craig, President's Report, 1902, 47–50, UniPub, S2, B1895–1912; also "Timber Harvesting," at http://www.foresthistory.org /ASPNET/Publications/region/1/flathead/chap10.htm.

28. Craig, President's Report, 1907, 25–33, UniPub, S2, B1895–1912.

29. Craig, President's Report, 1902, 6, UniPub, S2, B1895–1912; also RG1, PO, S4, Box 168, F "Applications"; William R. Trowbridge, Principal, Helena High School, to Dr. Craig, 31 July 1905, RG1, PO, S4, Box 168, F "Correspondence Received—Dr. O. J. Craig, about 1896–1907 . . ."; and "Table Showing the number of graduates from Montana High Schools that enter Higher

Education" (no date, but after 1899), RG1, PO, S4, Box 168, F "Letters Concerning Vocations. . . ."

30. Craig, President's Report, 1907, 5–7; and Clyde A. Duniway, President's Report, 1908, 9–10, both UniPub, S2, B1895–1912.

31. Craig, President's Report, 1902, 7; 1903, 22; and 1905, 22, all UniPub, S2, B1895–1912; also O. J. Craig, President, to C. F. Mellen, President Northern Pacific Railroad, 24 February 1902, RG1, PO, S4, B167, F "Correspondence—1901–1908." See also Scheuch, "University's Start Was Very Modest," *Missoulian, Souvenir Edition* (20 July 1922), 9–12, RG1, PO, S15, B26, F "ARCHIVES: Informational."

32. Morton J. Elrod, "First Football Team [1897], University of Montana" (no date, but 1920s), MJE, S5, B27, F11.

33. Craig, President's Report,1902, 9; and Craig, President's Report, 1903, 12, both UniPub, S2, B1895–1912.

34. Craig, President's Report, 1905, 24, UniPub, S2, B1895–1912.

35. Elrod, "Many Changes," *Missoulian* (29 January 1922), 1, 5, MJE, S3, B7, F14. Also Merriam, *History,* 2 and 182, for slightly different wording and capitalization: "The University—it shall prosper," and "The University of Montana—It Shall Prosper."

36. Morton J. Elrod, "Among the Kootenais," *Rocky Mountain Magazine* 3, nos. 3–4 (November–December 1901), 171–85, MJE, S3, B7, F11.

37. Morton J. Elrod, "Reminiscences," 18–20 January 1921, excerpt from *History of Montana* (no date), 878–82, MJE, S3, B5, F1.

38. Morton J. Elrod, Secretary-Treasurer, to Members, Montana Horticultural Society (no date, but 1905–1915), MJE, S2, B3, F14.

39. Elrod, "Department of Biology," in Craig, President's Report, 1903, 40–45, UniPub, S2, B1895–1912.

40. Morton J. Elrod, "Whence Comes the Water" (undated, but about 1910), MJE, S3, B7, F23.

41. Elrod, "Department of Biology," in Craig, President's Report, 1903, 40–45, UniPub, S2, B1895–1912.

42. Morton J. Elrod, "Report of the Department of Biology," 1903–1904, MJE, S5, B23, F8; Morton J. Elrod, "The Department of Biology," in Oscar J. Craig, President's Report, 1904, 43–47, UniPub, S2, B1895–1912; and the appendix herein.

43. Morton J. Elrod, "Department of Biology," in Craig, President's Report, 1905, 43–46; and Craig, President's Report, 1905, 25–28, both UniPub, S2, B1895–1912.

44. Elrod, "Department of Biology," in Craig, President's Report, 1903, 40–45, UniPub, S2, B1895–1912. Emphasis supplied in the quotation from Elrod to Furst, 1916, MJE, S2, B2, F2.

45. Morton J. Elrod, "Department of Biology," in Clyde A. Duniway, President's Report, 1908, 42–43; and Duniway, President's Report, 1908, 12–14, both UniPub, S2, B1895–1912.

46. Morton J. Elrod, "Department of Biology Report," 27 November 1910, MJE, S5, B23, F6; and Elrod, "Department of Biology," in Clyde A. Duniway, President's Report, 1910, 50–53, UniPub, S2, B1895–1912.

47. Clapp, "Narrative," ch. 3, 16–21; and Merriam, *History,* 34.

48. Craig, President's Report, 1903, 13–15, UniPub, S2, B1895–1912; and see SBE, Minutes, Regular Meeting, 6 June 1904, 504–508, M418, R2. Twenty-eight accredited high schools by 1908, see "At the Montana State University," *Missoulian* (30 August 1908), 1, 3, R32.

49. Craig, President's Report, 1907, 24–33, UniPub, S2, B1895–1912.

50. See an excerpt from the *Sentinel* (1924), Elrod "beloved by all," "a sympathetic listener, and an invaluable 'troubleshooter' for students," MJE, S5, B23, F2. Also 67 Alumni "To the Honorable President of the University of Montana" (no date but June–August 1908), and "We, the undersigned students," to the President (no date but June–August 1908), RG30, HR, F "Elrod, M. J."

51. See Excerpts from "Faculty Minutes," 15, 22 January and 5 February 1918, concerning the student council with Elrod as one of three elected faculty members, RG1, PO, S18, B21, F "Student Affairs—Student Council 1918"; "At the Montana State University," *Missoulian* (30 August 1908), 3, R32; and "Elrod, Morton J., 1863–1953," at http://socialarchive.iath.virginia.edu/xtf/view?docId =elrod-morton-j-1863-1953-cr.xml. See also Morton J. Elrod, "To the Faculty," 1 December 1927, RG1, PO, S4, B167, F "Campus Christmas Tree."

52. M. J. Elrod, W. M. Aber, and Eunice J. Hebbell, "To the Faculty of the University of Montana," 13 March 1898, MJE, S5, B27, F14; "The First Milestone," *Anaconda Standard* (30 May 1898); Morton J. Elrod, "How Shall a Vacation Be Spent," *Kaimin* (1898), 6–8, both MJE, S8, B37, F24; and Clapp, "Narrative," ch. 2, 34, and ch. 3, 25–26.

53. Giltner, "Montana State University," 15 February 1939, 4–5; and F. C. Scheuch, Acting President, "Data Concerning the State University," 1916, both RG1, PO, S15, B26, F "ARCHIVES: Informational." Also "Hiram Boardman Conibear," at http://en.wikipedia.org/wiki/Hiram_Boardman_Conibear; Ellen Garvin, "And in the Beginning . . ." (Senior Practice Laboratory, School of Journalism, The State University of The University of Montana, Missoula, Montana, 18 May 1925), 4–6, copy in author's possession; and J. P. Rowe, Chairman, Interscholastic Committee, to Morton J. Elrod, 23 January 1935, regret caused by "the first time you have been absent from the Interscholastic Committee during its entire life," MJE, S2, B3, F9. Also Morton J. Elrod to Clyde A. Duniway, 25 June 1908, raising funds to support students in the meet, CAD, MSS 735, R4.

54. "At the Montana State University," *Missoulian* (30 August 1908), 3, R32.

55. Craig, President's Report, 1905, 25–28, UniPub, S2, B1895–1912.

56. "From the faculty minutes," 10 January 1898; and L. K. Williams, "The Montana State University School of Pharmacy" (no date but 1927), no

pagination, both RG1, PO, S15, B37, F "Pharmacy: 1916–1932." See also Craig, President's Report, 1902, 19, UniPub, S2, B1895–1912.

57. "At the Montana State University," *Missoulian* (30 August 1908), 3, R32; and Nathaniel Craighill to Clyde A. Duniway, July 1908, CAD, MSS 735, R4.

58. J. H. T Ryman to Clyde A, Duniway, 26 June, 18 July, and 10 August 1908; William A. Aber to Clyde A. Duniway, 14 July 1908; W. E. Harman to Clyde A. Duniway, 19 June and 14 July 1908, all CAD, MSS 735, R4.

59. Morton J. Elrod, "The Museum," December 1901–November 1902, MJE, S5, B23, F6.

60. Elrod, untitled request, 1915–16, MJE, S5, B28, F6.

61. M. J. Elrod to F. C. Scheuch, 6 May 1916, RG1, PO, S15, B26, F "ARCHIVES: Museum, 1915–1944." For a description of the growing collections, see Elrod, *Biological Reconnaissance,* esp. 93–96, with photographs.

62. Elrod, untitled request, 1915–16, MJE, S5, B28, F6. In 1930, the State University valued the museum contents at $437,724, Charles H. Clapp, Report to the U.S. Bureau of Education, Department of the Interior, 1930, RG1, PO, S19, B46, F "Bureau of Education Reports and Letters—1916–1931."

63. Craig, President's Report, 1903, 15–17, UniPub, S2, B1895–1912; Oscar J. Craig, President's Report, 1904, 10–12, and Morton J. Elrod, "Department of Biology," in Craig, President's Report, 1904, 43–47, UniPub, S2, B1895–1912. Also W. B. Parsons to Morton J. Elrod, 15 June 1916, MJE, S2, B2, F2; and F. C. Scheuch to Dr. W. B. Parsons, 20 May 1916, all RG1, PO, S15, B36, F "Museum, 1915–1944."

64. M. J. Elrod to F. C. Scheuch, Acting President, 15 April and 16 July 1916; Mrs. H. E. Fearnell, Miles City, to Prof. M. J. Elrod, 14 June and 6 July 1915; and J. P. Rowe to F. Scheuch, 17 April 1916, all RG1, PO, S15, B36, F "Museum, 1915–1944."

65. See the letters cited in the preceding note and others in the same folder.

66. Oscar J. Craig, President's Report, 1898, 8, UniPub, S2, B1895–1912; also Morton J. Elrod, "Air Currents in Montana," 1904, MJE, S3, B7, F1; N. T. Gisborne, Division of Forest Management, Rocky Mountain Forest and Range Experiment Station, Missoula, to W. R. Krumm, U.S. Weather Bureau, Missoula, 29 November 1944, MJE, S2, B3, F10; Mary Elrod Ferguson, typescript summary of Elrod's life (no date, but 1950s), MJE, SI, B1, F1. See also Morton J. Elrod to Clyde A. Duniway, 25 June and 1 July 1908; Joseph M. Dixon to Clyde A. Duniway, 3 July 1908; and Clyde A. Duniway to Joseph M. Dixon, 11 July 1908, all CAD, MSS 735, R4. Also see "S. 5206," introduced by Dixon, 11 February 1908, RG1, PO, S4, B168, F "Bills, etc., 1898–1899 . . . 'Building Commission'—About 1905"; and Elrod, "Department of Biology," 1910, in Duniway, President's Report, 1910, 50–53; also Clyde A. Duniway, President's Report, 1908, 10–11, UniPub, S2, B1895–1912. See also "Dixon Brings Good News,"

and "Establishes Weather Station to be at the University," *Missoulian* (21 June and 16 August 1908), 1–12 and 1, respectively, R32. B. F. Young installed the equipment and W. D. Harkins served as observer during Elrod's absence while away at the Biological Station.

67. Elrod, *Biological Reconnaissance,* 91–96. Also Morton J. Elrod, "Biological Station," 29–31, in Craig, President's Report, 1899, UniPub, S2, B1895–1912; also Elrod, "Brief History," 1920, MJE, S5, B29, F5; and Beard, "Morton J. Elrod," Section 3, 1–13, MJE, S1, B1, F1.

68. O. J. Craig, President, "Recommendations," December 1899, and J. M. Hamilton and J. G. McKay, "Report of University Committee," December 1899, both RG1, PO, S5, B11, F "President—Craig, O. J. No. 4."

69. Morton J. Elrod, "The University of Montana Biological Station," 1910; Morton J. Elrod, "Biological Station, Outline for 1916," both MJE, S5, B23, F6; and Elrod, "Brief History," 1920, MJE, S5, B29, F5.

70. Elrod, "Biological Station," 29, in Craig, President's Report, 1899, UniPub, S2, B1895–1912.

71. Beard, "Morton J. Elrod," Section 3, 16, MJE, SI, B1, F1. And see Robert Oslund, Woods Hole, to Morton J. Elrod, 24 July 1921, MJE, S2, B2, F6; and Morton J. Elrod to Chancellor Edward C. Elliott, 2 March 1917, MJE, S5, B30, F2. Private conversation in 1993 with Dr. Jessie Bierman by the author.

72. Marcus E. Jones to President Duniway, 16 December 1908, MJE, S2, B1, F10; and "Who was Marcus E. Jones," at https://sites.google.com/a/rsabg .org/marcus-e-jones/home/who-is-marcus-e-jones.

73. "Flathead Lake Biological Station," at http://flbs.umt.edu/.

74. Elrod, "Brief History," 1920, MJE, S5, B29, F5.

75. Morton J. Elrod to President Clapp, 2 February 1933, MJE, S3, B5, F1; and Morton J. Elrod, "Biological Station at Flathead Lake" (no date but 1898), both MJE, S6, B34, F10.

76. William A. Clark to Morton J. Elrod, 26 June, 30 June, and 27 December 1899, $250 for 1900 as well, all MJE, S2, B1, F8.

77. Elrod, "Brief History," 1920, MJE, S5, B29, F5. Also Elrod, "Biological Station," 29, in Craig, President's Report, 1899, UniPub, S2, B1895–1912; and Morton J. Elrod to Clyde A. Duniway, 1 July 1908, CAD, MSS 735, R4. See also Morton J. Elrod, "Vacation in the Rocky Mountains" (no date but 1898–1899), MJE, S3, B7, F22; Elrod, *Biological Reconnaissance,* passim; and Beard, "Morton J. Elrod," Section 2, 9–10, MJE, SI, B1, F1.

78. Elrod, *Views of the Mission Mountains.* See also Elrod, "Among the Kootenais," *Rocky Mountain Magazine* 3, no. 3–4 (November–December 1901), 171–85, MJE, S3, B7, F11; and Morton J. Elrod, "The Beauties of the Mission Range" (1901), offprint from *Rocky Mountain Magazine* 2, no. 2 (April 1901), 623–31, MJE, S3, B7, F12.

79. See Elrod, *Biological Reconnaissance,* "The Buffalo Herd," 130–34.

80. American Bison Society, *Annual Report, 1905–1907* (np: American Bison Society, 1908), 1–14, at http://archive.org/stream/annualreportofambs00 amer#page/n29/mode/2up.

81. Diettert, *Grinnell's Glacier,* 47–48.

82. Morton J. Elrod, "The Flathead Buffalo Range: A Report to the American Bison Society . . . ," American Bison Society, *Annual Report, 1905–1907* (np: American Bison Society, 1908), at 15–49, MJE, S3, B4, F16; and Hornaday, "Report of the President," *Second Annual Report of the American Bison Society,* 1–19, mentioning the honorarium of "several hundred dollars," and referring to Elrod as an expert.

83. "George Bird Grinnell," at http://en.wikipedia.org/wiki/George_Bird _Grinnell.

84. "Senate Bill 6159—A Bill to establish a permanent national bison range," introduced by Senator J. M. Dixon, 16 March 1908, MJE, S3, B4, F1; later amended to provide 20,000 acres and another $3,000 to cover the cost of the fences, with $2,500 appropriated by the state to help with the purchase of bison, as see state Representative J. D. Garber to Senator Joseph M. Dixon, 9 February 1909; and Senator Joseph M. Dixon to Representative J. D. Garber, 13 February 1909, JMD, MSS 055, B10,F6; and for Dixon's role in passing the act, see Senator Joseph M. Dixon to Dr. William T. Hornaday, 21 May 1908, JMD, MSS 055, B9, F2; Hornaday, "Report of the President," *Second Annual Report,* 1–19; Karlin, *Dixon of Montana* esp. vol. 1, 54–61, and vol. 2, 242; exchanges between Hornaday and Dixon and Gifford Pinchot and President Theodore Roosevelt, mentioning Elrod's role, JMD, MSS 055, B8, F10–11 and B9, F2; and Smith, "Politics of Allotment," *Pacific Northwest Quarterly,* 130–40, on opening the Flathead Reservation. For relocation of the Flathead or Salish-Kootenai tribes from Ravalli County to the Flathead Reservation, see Stone, *Following Old Trails,* 83–96.

85. Elrod, "Buffalo Range," 17–48, MJE, S3, B4, F16; also Morton J. Elrod, "Biological Station," 61–64, in Craig, President's Report, 1907, UniPub, S2, B1895–1912; and Elrod, *Biological Reconnaissance,* 133.

86. Elrod, "Passing of the Pablo Buffalo Herd," 35–41; and Elrod, "The Montana Bison Range," 45–48, both MJE, S3, B4, F12. Also "Transported Buffalo," *Missoulian* (9 May 1910), R39, for a report concerning the effort to capture the obstinate animals and a comment by Alex Ayotte, the Canadian Dominion park inspector, commending the quality of the Montana range selected by Elrod.

87. Morton J. Elrod, handwritten on Pablo herd (undated but after 1910), MJE, S3, B4, F15. Charlie Russell to Bertrand W. "Fiddleback" Sinclair, January 1909, in *Your Friend, C. M. Russell: The C. M. Russell Museum Collection of Illustrated Letters,* Ann Morand (Great Falls, MT: C. M. Russell Museum, 2008), 33–41; and Jennifer Bottomly-O'Looney and Kirby Lambert, *Montana's Charlie Russell: Art in the Collection of the Montana Historical Society* (Helena: Montana

Historical Society, 2014), 332–33, on Russell's opportunity to see wild buffalo. Also Hornaday, "President's Report," passim, for the effort to solicit $10,000 from private donors to purchase forty wild buffalo, of which Elrod raised $312 in Montana. See also exchanges between Dixon and Hornaday on purchasing buffalo from either Pablo or the Conrad estate (Conrad had purchased animals from Pablo) with privately donated and state-appropriated money, JMD, MSS 055, B9, F2; and W. T. Hornaday to Senator Joseph M. Dixon, 30 September 1909, JMD, MSS 055, B11, F2, indicating purchase of thirty-four (twenty-two females) and gift of two more from the Conrad estate, gift of ten from Charles Goodnight of Texas, and another gift from J. J. Hill—all purebloods. Pablo refused to sell or donate. On the acquired buffalo, see H. K. C. Bryant, "The Jaunts of a Tenderfoot: III—On the Bison Range," *Missoulian* (21 July 1910, R39; and Elrod, *Biological Reconnaissance,* 130.

88. "Biological Station Ground," 1910, MJE, S5, B30, F3; President C. H. Clapp to Chancellor M. A. Brannon, 2 February 1933; and Morton J. Elrod to President Clapp, 2 February 1933, both MJE, S3, B5, F1. See also Elrod, "Brief History," 1920, MJE, S5, B29, F5; Morton J. Elrod, "Biological Station," 49, in Craig, President's Report, 1905, UniPub, S2, B1895–1912; and Craig, President's Report, 1906, 15, UniPub, S2, B1895–1912; Morton J. Elrod, "Biological Station," 45–46, in Craig, President's Report," 1906, UniPub, S2, B1895–1912; and Beard, "Morton J. Elrod," Section 4, 2–4.

89. John K. Rankin to O. J. Craig, President, 7, 11, 15 September 1906, RG1, PO, S4, B168, F "Correspondence—1901–1908."

90. Morton J. Elrod, "Biological Station," 61–64, in Craig, President's Report, 1907, S2, B1895–1912.

91. Elrod, "Brief History," 1920, MJE, S5, B29, F5.

92. Elrod, "First Story," in Elrod, "Four Stories about the Flathead" (no date, but 1908–1916), MJE, S3, B5, F8.

93. Clapp, "Narrative," ch. 2, 52.

94. Elrod, "Fourth Story," in Elrod, "Four Stories About the Flathead" (1908–16), MJE, S3, B5, F8.

95. Morton J. Elrod to Chancellor Edward C. Elliott, "Report . . . of the Biological Station," 1916, MJE, S5, B29, F10.

96. Morton J. Elrod, "Biological Station," 46–48, in Oscar J. Craig, President's Report, 1900, UniPub, S2, B1895–1912; Elrod, *Biological Reconnaissance,* passim; and Morton J. Elrod, "Biological Station," 45–47, in Oscar J. Craig, President's Report, 1901, UniPub, S2, B1895–1912.

97. Elrod, "Biological Station," 61–64, in Craig, President's Report, 1907; Elrod, "Biological Station," 29, in Craig, President's Report, 1899; and Craig, President's Report, 1900, 36–37, all UniPub, S2, B1895–1912. Also Elrod, "First Report, Biological Station," 1899, MJE, S5, B28, F19; Morton J. Elrod, "Report . . . Biological Station," 1902, MJE, S5, B28, F21; Morton J. Elrod, "Report of . . . Biological Station," 1904, MJE, S5, B28, F22; Morton J. Elrod, "Summer of 1908

. . . Biological Station," 1908, MJE, S5, B28, F24; Morton J. Elrod, "The Biological Station," 1913, MJE, S5, B28, F14; and Morton J. Elrod, "Flathead Lake," 1914, MJE, S5, B29, F5. Also folders on station work from 1899 to 1921, MJE, S5, B28–30.

98. "The History of Summer Term Idea," *Missoulian* (18 December 1927), 1, 10; "The Summer School of Science"; both RG1, PO, S7, B7, F "Summer Session in 1927."

99. Elrod, "Biological Station," 61–64, in Craig, President's Report, 1907, UniPub, S2, B1895–1912; and the appendix herein.

100. Morton J. Elrod, "The Physiography of the Flathead Lake Region," 1902, MJE, S5, B33, F1; Elrod, *Biological Reconnaissance,* "Flathead Lake," 135–43, and "On the Swan River," 153–55; and Morton J. Elrod, "The Summer of 1908 at the University of Montana Biological Station," 17 November 1908, RG1, PO, S15, B30, F "Biological Station Reports—1908, 1914, 1928–1929."

101. "Flathead River," at http://en.wikipedia.org/wiki/Flathead_River; and "Swan River (Montana)," at http://en.wikipedia.org/wiki/Swan_River_(Montana).

102. See the list of publications in the appendix herein; and Elrod, *Biological Reconnaissance,* "Bibliography," 174–76.

103. Elrod, "Biological Station," 29, in Craig, President's Report, 1899, UniPub, S2, B1895–1912.

104. "Entomostraca," in http://en.wikipedia.org/wiki/Entomostraca; Craig, President's Report, 1900, 36–37, UniPub, S2, B1895–1912; and Elrod, *Biological Reconnaissance,* 107–109, with a photograph of the canvas boat, the *Daphnia,* and a description of the hose and pump apparatus for securing water samples at various depths up to 140 feet.

105. Elrod, "First Report, Biological Station," 1899, MJE, S5, B28, F19; Elrod, "Report . . . Biological Station," 1902, MJE, S5, B28, F21; Elrod, "Report of . . . Biological Station," 1904, MJE, S5, B28, F22; Elrod, "Summer of 1908 . . . Biological Station," 1908, MJE, S5, B28, F24; Morton J. Elrod, "The Biological Station," 1913, MJE, S5, B28, F14; Elrod, "Flathead Lake," 1914, MJE, S5, B29, F5. See also Morton J. Elrod to Chancellor Edward C. Elliott, 7 June 1916, MJE, S5, B29, F9. Morton J. Elrod, typescript "List of Papers and articles submitted for publication" (no date but 1921), MJE, S5, B30, F12; and the composite list in the appendix herein.

106. Morton J. Elrod to Dr. H. B. Ward, 29 June 1916, MJE, S5, B29, F9; also "Henry B. Ward Papers, 1885–1960," at http://www.amphilsoc.org/mole/view?docId=ead/Mss.Ms.Coll.41-ead.xml;query=.

107. On "Copepods" at http://en.wikipedia.org/wiki/Copepod#Diet; Morton J. Elrod, "Flathead Lake Report," 1916, MJE, S5, B30, F1; and Morton J. Elrod, "Parasitic Copepoda Found on Fins and Gills of Bull Trout," *Proceedings of the United States Museum* 47 (1915), MJE, S6, B30, F12.

108. Morton J. Elrod, "Recommendations" (no date but 1930), MJE, S5, B32, F10.

Chapter 2

1. The Company built the smelter in 1903, *Anaconda Standard, Butte Miner, Butte Inter Mountain,* and *Helena Independent Record,* "Smoke Fumes" Scrap Book, MSS 753, Scrapbook 4, K. Ross Toole Archives, Maureen and Mike Mansfield Library, The University of Montana; also Donald MacMillan, "A History of the Struggle to Abate Air Pollution From Copper Smelters of the Far West, 1885–1933" (PhD diss., University of Montana, 1973), and *Smoke Wars* (Helena: Montana Historical Society Press, 2000); "Full text," at http://www.23 .us.archive.org/stream/cu31924051103566/cu31924051103566.djvu.text.

2. Elrod, "Report of the Department of Biology," 1903–1904, MJE, S5, B23, F8; Elrod, untitled request, 1915–16, MJE, S5, B28, F6; Morton J. Elrod, "Answers to Inquiries of State Superintendent H. A. Davee," 23 May 1912, MJE, S5, B23, F1; Morton J. Elrod "Photography in the Park" (no date but file marked 1912), MJE, S4, B18, F12; and Morton J. Elrod, "Photography in the Mountains" (1918), MJE, S3, B7, F19. See also the composite list of publications in the appendix herein.

3. Judge's decision, Court of Appeals of the United States, "Bliss v. Washoe Copper Co.," in *Federal Reporter* 186 (St. Paul: West Publishing, 1911), 789–828. Available at https://books.google.com/books?id=HyQ4AAAAIAAJ&lpg =PA801&dq=Bliss%20v.Washoe%20Copper%20Smelter&pg=PA801#v=one page&q=Bliss%20v.Washoe%20Copper%20Smelter&f=false.

4. See nineteen photographs of 1905, MJE, S9, B71, F141–59; and a photograph of Harkins, MJE, S9, B48, F49.

5. "Smoke Not Wholly to Blame . . . ," "Did Not Consider Alkali . . . ," and "Would not Venture An Opinion . . . ," *Butte Inter Mountain* (3, 4, 5 April 1906), 1, 8; 1, 5; and 1, 10, respectively.

6. Morton J. Elrod, handwritten "report" (June 1906), MJE, S8, B37, F21; and notes, MJE, S7, B35, F17.

7. MacMillan, "Struggle to Abate Air Pollution," 154; Tulli Kerstetter, "J. W. Blankinship," at http://www.montana.edu/mlavin/herb/jwb.htm,

8. The judge reserved the right of all affected landowners to sue for damages if they had proof. Judge's decision, Court of Appeals of the United States, "Bliss v. Washoe Copper Co.," in *Federal Reporter* 186 (St. Paul: West Publishing, 1911), 789–828. Available at https://books.google.com/books?id=HyQ4 AAAAIAAJ&lpg=PA801&dq=Bliss%20v.Washoe%20Copper%20Smelter&pg =PA801#v=onepage&q=Bliss%20v.Washoe%20Copper%20Smelter&f=false.; and MacMillan, "History of the Struggle," 160. For plaintiff's attempted arbitration after the adverse decision, see also "Smoke Fumes," MSS 753, 122–28, 171, 175, 222–23, 234–36; Rice and Kauffman, "William Draper Harkins," *Bulletin for the History of Chemistry* 20 (1997), 60–67, at http://www.scs.illinois. edu/~mainzv/HIST/bulletin_open_access/num20/num20%20p60–67.pdf. After 1908, the federal government sought to force the Anaconda Company to

mitigate the toxic effects, but with little success. The company purchased land around the smelter from owners willing to sell, and ultimately traded sections of timberland elsewhere for affected state timberland around the smelter.

9. Craig, President's Report, 1906, 11–15, 24, UniPub, S2, B1895–1912.

10. William Draper Harkins, "Chemistry Report," Craig, President's Report, 1906, 40, UniPub, S2, B1895–1912.

11. Elrod, "Biological Station Report," Craig, President's Report, 1906, 45–46, UniPub, S2, B1895–1912.

12. Ibid.; Morton J. Elrod to Emma, 2 July 1901, MJE, S2, B1, F9. For a detailed description of the 1906 excursion, see Elrod, "The Proposed Glacier National Park" (probably 1910), MJE, S4, B18, F15; Morton J. Elrod, "The attempt on St. Nicholas" (no date but probably 1906), MJE, B19, F9–11, handwritten in 148 pages and 55 typed pages.

13. Morton J. Elrod, "Introduction," handwritten (no date but 1926–30), trips in 1906, 1909, 1910, 1911, and 1914, in file titled "Early Expeditions to Glacier National Park, 1906–1914," MJE, S4, B15, F12; Morton J, Elrod, handwritten notes on the park (no dates but 1914–20), MJE, S4, B18, F7; Morton J. Elrod, "The University of Montana Biological Station," 1910, MJE, S5, B28, F25. Also Morton J. Elrod, "A Study of Timber," 1906, MJE, S4, B15, F5; Morton J. Elrod, "The Garden Wall in Glacier National Park Presents a Scene of Great Beauty Unexcelled," *Missoulian* (25 June 1911), 3, MJE, S4, B15, F21; Morton J. Elrod, "Trip to Sperry Glacier" (no date but 1931); and Morton J. Elrod, "Sperry (Comeau) Pass" (no date but 1931), both MJE, S4, B20, F1. The "Introduction" dated the first excursion in 1908, but Elrod's station report for 1906 and his discussions in "The Proposed Glacier National Park" (probably 1910), MJE S4, B18, F15, and "Attempt on St. Nicholas," MJE, S4, B19, F9–11, gave details for the 1906 trip.

14. Craig, President's Report, 1907, 13, 24–25, 35, UniPub, S2, B1895–1912.

15. Morton J. Elrod, "Department of Biology," and "Biological Station Report," in Craig, President's Report, 1907, 60–64, UniPub, S2, B1895–1912.

16. SBE, Minutes, Regular Meeting, 3 June 1907, 85–90, M418, R2; "At the Montana State University," *Missoulian* (30 August 1908), 3, R32; and O. J. Craig to Mr. Orville Bremer, 6 January 1908, RG1, PO, S4, B167, F "Correspondence—1901–1908." See also President O. J. Craig to Senator J. M. Dixon, 10 July 1908, concerning Craig's search for a pension, JMD, MSS 055, B9, F2.

17. Clapp, "Narrative," ch. 3, 6–8.

18. Merriam, *History,* 18.

19. See Mrs. Tyler Thompson's comments, "Resolutions With an Amendment Is the Result of Craighead Meeting Influenced at Intervals by Hysteria," *Missoulian* (10 July 1915), 1–4, 8, MJE, S5, B26, F17; Miles Romney, "From An Independent Newspaper," *New Northwest* (no date but September 1921), reprinted from *Western News,* MJE, S5, B27, F10. The campus mythology describing a "graveyard of presidents" gained a life of its own.

20. W. E. Harmon to Clyde A. Duniway, 19 June, 14 July, and 8 August 1908, all CAD, MSS 735, R4.

21. J. H. T. Ryman to C. A. Duniway, 26 June, 18 July, and 10 August 1908; Clyde A. Duniway to J. H. T. Ryman, 3 and 21 July 1908, all CAD, MSS 735, R4.

22. SBE, Minutes, Regular Meeting, 3 June 1907, 85–90, M418, R2; see also SBE, Minutes Special Meeting, 1 April 1908, 92, M418, R2.

23. See several letters from principals, superintendents, scientists, higher education faculty and staff members, and college and university presidents, MJE, S5, B27, F2.

24. Charles C. Adams, University of Chicago, to Superintendent W. E. Harmon, 1 May 1908, MJE, S5, B27, F2.

25. Clapp, "Narrative," ch. 3, 6–8; SBE, Minutes, Regular Meeting, 2 December 1908, 88, M418, R2.

26. For no mention of the dispute, see Elrod, "The Biological Station," in Craig, President's Report, 1907, 61–64, UniPub, S2, B1895–1912; and Clyde A. Duniway, President's Report, 1908, 1–25, UniPub, S2, B1895–1912.

27. F. C. Scheuch, "Data Concerning the State University," 1916, no pagination, in RG1, PO, S15, B26, F "ARCHIVES: Informational."

28. SBE, Minutes, Regular Meeting, 2 December 1907, 88; and Minutes, Special Meeting, 1 April 1908, 96, both M418, R2.

29. Morton J. Elrod to Governor E. L. Norris, 29 March 1908, in SBE, Minutes, Special Meeting, 1 April 1908, 96, M418, R2. See also "Biological Station to Continue," *Missoulian* (8 July 1908), 3, R32; and "Biological Station Described," *Missoulian* (15 August 1908), quoting first issue of *Flathead County News.* No mention of Elrod's predicament in either article.

30. See "Biological Station to Continue," *Missoulian* (8 July 1908), R33, specifically discussing use of the Clark funds and listing the names of the visiting scientists.

31. SBE, Minutes, Regular Meeting, 1 June 1908, 98–105, M418, R2.

32. See Clapp, "Narrative," ch. 2, 6–8,using but altering Aber's meaning, as see Aber to Duniway, 13 July 1908, CAD, MSS 735, R4.

33. Aber to Duniway, 13 July 1908, CAD, MSS 735, R4. Clapp also misquoted this letter.

34. Ibid.

35. Harmon to Duniway, 14 July 1908, CAD, MSS 735, R4.

36. J. H. T. Ryman to C. A. Duniway, 26 June, 18 July, and 10 August 1908; and Clyde A. Duniway to J. H. T. Ryman, 3 July and 21 July 1908, all CAD, MSS 735, R4.

37. Clyde A. Duniway to J. B. Speer, 5 August 1908; and J. B. Speer to Clyde A. Duniway, 10 August 1908, both CAD, MSS 735, R5. For Speer's administrative philosophy, see J. B. Speer, "A Bird's-Eye View of the Organization of One University," *Journal of Higher Education* 2, no. 9 (December 1933), 461–67, RG1, PO, S19, B46, F "Questionnaires."

38. Merriam, *History,* 23, footnote 6.

39. See personnel documents, RG30, HR, F "Harkins, PROF. W. D."; also Clyde A. Duniway, President's Report, 1912, 6–7, reporting Harkins' resignation, UniPub, S2, B1895–1912. Harkins later won international distinction as a nuclear chemist, a field unrelated to his work as an expert witness in the smelter fumes litigation.

40. James R. Steele (apocryphal), "Hire Learning in Montana," *Pacific Weekly* (16 March 1936), excerpt, RG1, PO, S19, B167, F "Borton-Gosman Report 1939," for the claim of Elrod's dismissal; and for repetition of the charge, see Gutfeld, *Montana's Agony,* 117–18, citing the American Federation of Teachers, "The Keeney case: big business, higher education, and organized labor, report of an investigation," 1939, at http://www.worldcat.org/title/keeney-case -big-business-higher-education-and-organized-labor-report-of-an-investigation /oclc/13491059. Campus mythology had become accepted fact by 1939.

41. Jim Habeck, "Can Botanists Be Bought?" *Kelseya* (Summer 2006), at http://www.mtnativeplants.org/filelib/75.pdf.

42. See 67 Alumni "To the Honorable President of the University of Montana" (no date but June- August 1908); and "We, the undersigned [22] students," to the President (no date but June-August 1908), in RG30, HR, F "Elrod, M. J."

43. Morton J. Elrod to Clyde A. Duniway, 11 June 1908, CAD, MSS 735, R4.

44. "Dr. O. J. Craig Honored," *Missoulian* (4 June 1908), R32; Morton J. Elrod to Clyde A. Duniway, 25 June 1908, CAD, MSS 735, R4. For Elrod's later reconciliation with Craig's memory, see Elrod, "Many Changes," *Missoulian* (29 January 1922), 1, 5, MJE, S3, B7, F14; and Elrod, "Among the Kootenais," *Rocky Mountain Magazine,* 171–85, MJE, S3, B7, F11, for the naming of Craig Mountain in the Mission Range.

45. Morton J. Elrod to Clyde A. Duniway, 6 July 1908, CAD, MSS 735, R4.

46. Morton J. Elrod to Clyde A. Duniway (no date, but early July 1908, as see Clyde A. Duniway to Morton G. [*sic*] Elrod, 11 July 1908), receipt of the application, both CAD, MSS 735, R4. See also Arthur L. Stone, Editor, *Missoulian,* to President Clyde A. Duniway, 6 July 1908, CAD, MSS 735, R5; and William A. Aber to Clyde A. Duniway, 5 August 1908, CAD, MSS 735, R4.

47. Draft of Clyde A. Duniway to Morton J. Elrod, 19 June 1908, certain statements scratched out, CAD, MSS 735, R4. See also Clyde A. Duniway to W. D. Harkins, 6 July 1908, CAD, MSS 735, R4.

48. Duniway to Elrod, 19 June 1908, CAD, MSS 735, R4. For supportive letters Elrod said he did not solicit, see Senator Joseph M. Dixon to Dr. Clyde A. Duniway, 8 July 1908, urging Elrod's reinstatement and noting his excellent record; and Clyde A. Duniway to Senator Joseph M. Dixon, 16 July 1908, indicating his favorable opinion of Elrod and the high praise of Elrod by scientists, both JMD, MSS 055, B9, F2–3. See also Clyde A. Duniway to Henry B. Ward, 6 August 1908; Henry B. Ward, Dean of the University of Nebraska School of

Medicine, to Clyde A. Duniway, 30 July and 12 August 1908; Clyde A. Duniway to William A. Aber, 6 August 1908; William A. Aber to Clyde Duniway, 28 July 1908; and W. C. Staton, to Clyde A. Duniway, 13 July and 4 August 1908, all CAD, MSS 735, R4; and Stone to Duniway, 6 July 1908, mentioning Elrod had shown him a letter from Duniway, CAD, MSS 735, R5.

49. Draft of Duniway to Elrod, 19 June 1908, CAD, MSS 735, R4.

50. Clyde A. Duniway to Morton J. Elrod, 21 July 1908; also Morton J. Elrod to Clyde A. Duniway, 1, 6, and 19 July 1908, all CAD, MSS 735, R4. And see Morton J. Elrod to Emma and Mary, 21 July 1908, MJE, S2, B1, F10.

51. J. M. Evans to Clyde A. Duniway, 18 July 1908, CAD, MSS 735, R4.

52. J. D. Largent to Clyde A. Duniway, 13 July 1908, CAD, MSS 735, R4.

53. Clyde A. Duniway to J. M. Evans, 9 July 1908, CAD, MSS 735, R4.

54. J. M. Evans to Clyde A. Duniway, 14 July 1908, CAD, MSS 735, R4. However, Harmon stated earlier that the board had overruled the state university committee to deny Craig's reappointment.

55. G. T. Paul to Clyde A. Duniway, 20 July 1908, CAD, MSS 735, R4.

56. William A. Aber to Clyde A. Duniway, 5 August 1908, CAD, MSS 735, R4, emphasis supplied.

57. Ryman to Duniway, 26 June, 18 July, and 10 August 1908; and Clyde A. Duniway to J. H. T. Ryman, 3 July and 21 July 1908, all CAD, MSS 735, R4. Ryman knew the details of Craig's requested "retirement."

58. Clyde A. Duniway to W. E. Harmon, 10 August 1908, CAD, MSS 735, R4.

59. Clyde A. Duniway to E. L. Norris, 5 August 1908, CAD, MSS 735, R4.

60. Clyde A. Duniway to E. L. Norris, 17 July 1908; and E. L. Norris to Clyde A. Duniway, 29 July 1908, both CAD, MSS 735, R4.

61. Clyde A. Duniway to C. R. Leonard, 23 July, 2 August, and 11 August 1908, all CAD, MSS 735, R4.

62. C. A. Duniway to University Committee, State Board of Education, 24 August 1908, RG30, HR, F "Elrod, M. J."; see also "Affairs of the State University Discussed," *Missoulian* (24 August 1908), 1, R32.

63. "Dr. Morton J. Elrod is Re-Elected," *Missoulian* (15 September 1908), 1, R33, the first notice of Elrod's plight in the *Missoulian;* and SBE, Minutes, Special Meeting, 14 September 1908, 107–109, M418, R2, copy of original board notes in author's possession.

64. Elrod to Emma and Mary, 21 July 1908, MJE, S2, B1, F10, mentioning Emma had sent Duniway's letter of 25 July and Aber's of 9 July, the latter not in the Elrod papers.

65. C. A. Duniway, Stanford University, to Morton J. Elrod, 25 July 1908, CAD, MSS 735, R4. Clapp, "Narrative," ch. 3, 8, "after gathering still more evidence" Duniway "requested re-instatement," but Duniway decided much earlier to insist on it.

66. Elrod to Emma and Mary, 21 July 1908, MJE, S2, B1, F10.

67. Morton J. Elrod to Clyde A. Duniway, 5, 23, and 24 August 1908, all CAD, MSS 735, R4.

68. Morton J. Elrod, "The Summer of 1908 at the University of Montana Biological Station," 17 November 1908, RG1, PO, S15, B30, F "Biological Station Reports—1908, 1914, 1928–1929"; and Morton J. Elrod, "Biological Station," 42–45, in Duniway, President's Report, 1908, UniPub, S2, B1895–1912. Sliter allowed use of his land until 1909.

69. Elrod, *Views of the Mission Mountains.*

70. Elrod, "Summer of 1908," 17 November 1908, RG1, PO, S15, B30, F "Biological Station Reports—1908, 1914, 1928–1929."

71. Elrod, "Where the Red Man Wrote His Story," *Missoulian* (1 March 1908); Elrod, "Bulletins of the University" (dated 1908 in Morton J. Elrod to President E. O. Sisson, 22 January 1918, MJE, S5, B30, F4), MJE, S2, B1, F10.

72. Elrod to Duniway, 24 August 1908, CAD, MSS 735, R4.

73. "Dr. Morton J. Elrod is Re-Elected," *Missoulian* (15 September 1908), 1, R33; also SBE, Minutes, Special Meeting, 14 September 1908, 108–109, M418, R2.

74. Elrod, "Many Changes," for "vastly improving the standard of the institution . . . from partial to complete higher education." MJE, S3, B7, F14.

75. Duniway, President's Report, 1908, 6–75, UniPub, S2, B1895–1912. See also C. A. Duniway, President, to President J. M. Hamilton, 2 December 1908; and J. M. Hamilton, President, to President C. A. Duniway, 3 December 1908, both RG1, PO, S3, B16, F "1945 Legislative Data—University System"; and President Clyde A. Duniway to President C. H. Bowman, 2 December 1908, RG1, PO, S15, B33, F "Forestry and Conservation Experiment Station—through 1955."

76. On that debate, see *Weekly Missoulian* (various January–February 1893), R Jan. 4 1893–Dec. 26, 1894; Giltner, "Montana State University," 15 February, 1939, esp. 1–5; also Beatty, "Montana University's Birth Saw Big Debate, Lobbying," *Kaimin* (8 April 1957), 3, 5; and Scheuch, "University's Start Was Very Modest," *Missoulian, Souvenir Edition* (20 July 1922), passim, esp. 10–11, all RG1, PO, S15, B26, F "ARCHIVES: Informational."

77. For chartering acts, see Office of the Chancellor, "The University Code: Part I. I. Federal Statutes, Rulings, Regulations and Instructions II. State Constitutional Provisions III. State Statutes," *University of Montana Bulletin* (General Series, no. 10, June 1919), 190; and "Index," at 91, 93, and 101, RG1, PO, S4, B63, F "University Code."

78. SBE, Minutes, Regular Meeting, 7 December 1908, 115, M418, R2; and Clyde A. Duniway to E. O. Busenbury, Lewistown, 16 December 1908; and E. O. Busenbury to Dr. C. A. Duniway, 23 December 1908, both RG1, PO, S15, B26, F "ARCHIVES: Duniway Inauguration."

79. SBE, Minutes, Regular Meeting, 5 June 1911, 187, and Regular Meeting, 3 June 1912, 213, 216–17, M418, R2.

80. SBE, Minutes, Regular Meeting , 7 June 1909, 121–22 and 130, M418, R2. See also Clapp, "Narrative," ch. 3, 11; and Merriam, *History,* 27–28. For the local executive boards, see Office of the Chancellor, "The University Code," 1919, 135–41, at 137–39, Act of 1909, RG1, PO, S4, B63, F "University Code"; Speer, "A Bird's-Eye View," *Journal of Higher Education* 2, no. 9 (December 1933), 461–67, RG1, PO, S19, B46, F "Questionnaires."

81. Clapp, "Narrative," ch. 3, 12, from an unidentified Duniway letter to Senator Joseph M. Dixon, not in the Duniway papers.

82. Duniway, "Selections from . . . December 1909," 33–34, in Duniway, President's Report, 1910, UniPub, S2, B1895–1912; Merriam, *History,* 27, footnotes 15–16, quoting Duniway but stating Duniway suppressed the criticism that nonetheless appeared in the official board records.

83. Duniway, "President's Report, 1908," 5–6 and 18; Clyde A. Duniway, "Selections from . . . [unprinted report for] December 1909," in President's Report, 1910, 34–37; Duniway, President's Report, 1910; Duniway, President's Report, 1912, all UniPub, S2, B1895–1912. See also Merriam, *History,* ch. 2; Clapp, "Narrative," ch. 3; Clyde A. Duniway to President E. A. Bryan, WSC, 8 October 1908, RG1, PO, S15, B26, F "ARCHIVES: Duniway Inauguration"; Office of the Chancellor, "The University Code," at 167–68, Act of 3 March 1919, RG1, PO, S4, B63, F "University Code"; and "Regulations Governing Incidental Fee. Approved by the President, September 11, 1912" (no date but 1913–14), RG1, PO, S5, B11, F "President—Reports, 1913–1915, Folder 2."

84. Duniway, President's Report, 1908, 10–11, UniPub, S2, B1895–1912.

85. Merriam, *History,* 27–31; Clapp, "Narrative," ch. 3, esp. 55–59; and "At the Montana State University," *Missoulian* (30 August 1908), R32.

86. SBE, Minutes, Special Meeting, 31 July–1 August 1911, 190–93; and Regular Meeting, 4 December 1911, 194–95, M418, R2. See also Charles A. Kofoid, Chairman, "Academic Freedom and Academic Tenure: Report of the Committee of Inquiry Concerning Charges of Violation of Academic Freedom, Involving the Dismissal of the President and Three Members of the Faculty at The University of Montana," *Bulletin of the American Association of University Professors* 3, no. 5, part 2 (May 1917), 8–9, quoting a Duniway letter.

87. W. E. Harmon, Secretary, Executive Session of the Board, to President C. A. Duniway, 5 December 1911; and "From the Faculty Minutes of December 7, 1954," for Duniway's comments, both RG1, PO, B33, F "Forestry and Conservation Experiment Station—through 1955." See also Clyde A. Duniway to Henry S. Pritchett, 30 November 1912; Clyde A. Duniway to Professor M. V. O'Shea of Wisconsin, 2 February 1912; David Starr Jordan to Clyde A. Duniway, 22 January 1912; Clyde A. Duniway to W. E. Harmon, 26 April 1912; Clyde A. Duniway to President F. P. Keppel of Columbia, 24 January 1912; Arthur O. Lovejoy to Clyde A. Duniway, 2 February 1912, all CAD, MSS 735, R4; also Clyde A. Duniway to A. Lawrence Lowell, 19 December 1911, CAD, MSS 735, R5; and Clyde A. Duniway to Charles H. Hall, the member from

Missoula, with a handwritten note at the bottom indicating he had not sent it to avoid further controversy, both CAD, MSS 735, R5.

88. Merriam, *History*, 29–30, quoting Hall's statements; Hall later explained that Governor Norris charged the university committee to conduct a review of Duniway's performance and denied any personal hostility, "State Board Reaffirms Previous Action as to President Duniway," *Missoulian* (2 April 1912), 1, 7, MHS DN.

89. Merriam, *History*, 30. See also "Review of Duniway's performance, 1908–1911, and what occurred late that year," anonymous (no date, but late 1911 or early 1912), handwritten note at the top identifying Aber as the author, CAD, MSS 735, R5.

90. Jordan to Duniway, 22 January 1912, CAD, MSS 735, R4. Gutfeld, *Montana's Agony*, 118, asserted without evidence that the board fired Duniway because of political pressure.

91. Clyde A. Duniway, President's Report, 1912, 14–20, UniPub, S2, B1895–1912; discussed in attachment to Ernest O. Melby to Commission on Higher Education, 19 May 1944, RG1, PO, S19, B52, F "Duplication—Melby Attempts to Change Unit Functions, 1944–5."

92. See generally Merriam, *History*, ch. 3, 8, and 10; the definitive history not yet written.

93. Morton J. Elrod to Clyde A. Duniway, 22 December 1911, 29 April and 29 June 1912, all CAD, MSS 735, R5.

94. "Dr. Clyde Augustus Duniway," at http://www.findagrave.com/cgi-bin /fg.cgi?page=gr&GRid=87594794.

95. Jules A. Karlin, "Conflict and Crisis in University Politics," *Montana: The Magazine of Western History* 36, no. 2 (Summer 1985), 48–61; Clapp, "Narrative," ch. 4; and Merriam, *History*, ch. 3. Edwin B. Craighead to Governor E. L. Norris, 31 January 1913; and Edwin B. Craighead to Mr. Henry S. Pritchett, Carnegie Foundation, 10 January 1913, both RG30, HR, F "Craighead, Edwin B."

96. Clapp, "Narrative," ch. 4, and Merriam, *History*, ch. 3; "An Act for the Consolidation of the State Institutions of Higher Education" (no date but early 1913), with a handwritten note to Duniway; see also J. H. T. Ryman to Clyde A. Duniway, 18 January 1913, "Threshed Craighead plan today noonday lunch adjourning four thirty"; and Clyde A. Duniway to William A. Aber, 15 February 1913, asking about "the next move" with the failure of the "consolidation" bill, all CAD, MSS 735, R5.

97. Walter J. Greenleaf, U.S. Office of Education, "The Land-Grant College as It Functions To-Day," *School and Society* 41 (29 June 1935), 855–59, RG1, PO, S19, B46, F "Land-Grant Colleges." Concern about this issue influenced the original debate about consolidation in the 1890s, as see "Pointers for Consolidation" (25 January 1893); and "Practical Results Wanted," and "The Bunching Plan" (1 February 1893), all *Weekly Missoulian*, R Jan. 4, 1893–Dec. 26, 1894.

See also "Splendid Progress at the State University Emphasized By Commencement and Its Events," *Missoulian* (15 June 1910), R39, concerning the unnecessary duplication and lowering of standards by the agriculture schools—even in Montana—essentially to attract students, "will adjust" themselves automatically in time.

98. See excerpt of notice attached to Edwin L. Norris to Doctor E. B. Craighead, 22 January 1913, RG30, HR, F "Craighead, Edwin B."

99. SBE, Minutes, Regular Meeting, 2 and 23 December 1912, 224, 230, M418, R2; and Merriam, *History*, 37. For Hamilton to Duniway, 24 January 1909, and 3 December 1908, RG1, PO, S3, B16, F "1945 Legislative Data—University System."

100. Luke Wright, "Education Crisis on Anniversary of Others," *Great Falls Tribune* (9 June 1963), 1, 20, RG1, PO, S2, Box 25, F "ARCHIVES, History"; Paris Gibson, Chairman, Consolidation Committee, "Consolidation Handbook" (Missoula: Consolidation Committee, no date, but 1913–14), 30; W. M. Aber, "Consolidation vs. Segregation: 'Come Let Us Reason Together'" (Pamphlet, University of Montana, 11 February 1913), 4; and "Plan for Creation of A Greater University" (no place or date but 1913), 2, all RG1, PO, S4, B167, F "Consolidation." Also see numerous press excerpts and other materials, MJE, S5, B26, F16–19; and H. A. Davee, Superintendent, to "Teachers and University Alumni of Montana," 14 October 1914, MJE, S5, B26, F16.

101. E. B. Craighead, President, to Miss Gertrude Buckhous, 28 August 1914, RG1, PO, S4, B167, F "Consolidation."

102. "Governor Pledges Support in Eliminating Friction Between State's Colleges," *Missoulian* (13 January 1914), 1, 10, MJE, S5, B26, F17; and Sam V. Stewart, "Declaration," *Missoulian* (25 May 1914), MJE, S5, B26, F17. See also Paris Gibson, "Governor's Letter Should Be Studied," *Missoulian* (1 June 1914), 3, MJE, S5, B26, F7.

103. On the Leighton Act, see C. S. Stewart, Secretary of State, "Certificate," of "Senate Bill No. 105," 14 March 1913, RG1, PO, S15, B26, F "History of MSU." Also see Office of the Chancellor, "The University Code," June 1919, at 141*ff.*, Act of 14 March 1913, RG1, PO, S4, B63, F "University Code"; and Scheuch, "University's Start Was Very Modest," *Missoulian, Souvenir Edition* (20 July 1922), 9–12, esp. 11, RG1, PO, S15, B26, F "ARCHIVES: Informational."

104. Morton J. Elrod, "Montana State Board of Education Acts on Duplication of Courses," *Inter-Mountain Educator* 9, no. 1 (September 1913), 26–27; and "Engineering Department Transferred to Bozeman," *Missoulian* (18 July 1913), MJE, S5, B26, F17; also SBE, Minutes, Regular Meeting, 2, June 1913, 5, 6, 7, 14; Special Meeting, 17 July 1913, 23–26; and Regular Meeting, 22 December 1913, 35, 38, all M418, R2.

105. SBE, Minutes, Regular Meeting, 1 June 1914, 57, M418, R2. See also Merriam, *History*, 183, for the University charter prohibiting "instruction, either sectarian or partisan, in politics."

106. Stewart, "Declaration," *Missoulian* (25 May 1914), MJE, S5, B26, F17.

107. Kofoid, "Academic Freedom and Academic Tenure," *Bulletin of the American Association of University Professors* (May 1917), 12–13.

108. Ibid. Also comments by J. M. Dixon, "Resolutions With an Amendment is the Result of Craighead Meeting Influenced at Intervals by Hysteria," *Missoulian* (10 July 1915), 1–4, 8, MJE, S5, B26, F17; and Katlin, "Conflict and Crisis," esp. 58.

109. Merriam, *History,* 37–38; Karlin, "Conflict and Crisis," 46–48; and Clapp, "Narrative," ch. 4, 8–9. See also SBE, Minutes, Special Meeting, 11 October 1915, 93–95, on Chancellor Edward C. Elliott, M418, R2.

110. J. H. T. Ryman, "A Statement," *Missoulian* (18 July 1915), MJE, S5, B26, F20; J. H. T. Ryman to Morton J. Elrod, 30 July 1917, MJE, S2, B2, F3; SBE, Minutes, Regular Meeting, 7 June 1915, 83, 85, 88–89, M418, R2; "Board Refuses to Renew Craighead's Contract; Hamilton Resigns from State College Presidency," *Missoulian* (8 June 1915), MJE, S6, B34, F16; and Kofoid, "Academic Freedom and Academic Tenure," *Bulletin of the American Association of University Professors* (May 1917), esp. 16–30. Gutfeld, *Montana's Agony,* 118–19, claimed the Board fired Craighead rather than refusing to renew his contract.

111. Merriam, *History,* 37–38, 41–42; and "Board Refuses to Renew Craighead's Contract; Hamilton Resigns from State College Presidency," *Missoulian* (8 June 1915), MJE, S6, B34, F16. Also SBE, Minutes, Regular Meeting, 7 June 1915, 79–89; Minutes, Special Meeting, 28 April 1916, 114–15; and Minutes, Regular Meeting, June 1918, 148–49, M418, R2. Also "Should Adopt Amendment," *Butte Miner* (8 November 1914); and "A Word About Consolidation," *Missoulian* (27 November 1914), both MJE, S5, B26, F17. Also Karlin, "Conflict and Crisis," 48–61.

112. Edwin B. Craighead, "To the Editor" (no date, but September 1915), typescript sent to *Missoulian* and other papers, RG30, HR, F "Craighead, Edwin B."; Clapp, "Narrative," ch. 4–5; and Merriam, *History,* ch. 3. Merriam erroneously identified South rather than North Dakota.

113. Clapp, "Narrative," ch. 5; and "Through Personal Revenge Professors Are Dismissed," *Missoulian* (9 July 1915), MJE, S5, B26, F17.

114. Mary Stewart, Dean of Women, to President E. B. Craighead, 17 October 1914, RG30, HR, F "Craighead, Edwin B."; Karlin, "Conflict and Crisis," 48–61; and Merriam, *History,* 24, note 1, for the suggestion that Stewart's "independence brought on unpleasant rumors," confirmed by Karlin about her relationship with Ryman.

115. Kofoid, "Academic Freedom and Academic Tenure," *Bulletin of the American Association of University Professors* (May 1917), 3–52.

116. Karlin, "Conflict and Crisis," 48–61; and A. N. Whitlock to J. B. Speer, Registrar, 29 December 1927, RG1, PO, S5, B67, F "Ryman Scholarship."

117. Dixon's comments, "Resolutions With an Amendment is the Result of Craighead Meeting Influenced at Intervals by Hysteria," *Missoulian* (10 July 1915), 1–4, 8, MJE, S5, B26, F17.

118. Morton J. Elrod, "Consolidation of the State Institutions," *Inter-Mountain Educator* 9, no. 9 (May 1914), 27–30; "Consolidation of the State Institutions," 9, no. 10 (June 1914), 18–21, 24–25; "Consolidation of State Institutions," 10, no. 1 (September 1914), 25–26; "Consolidation of the University, the Agricultural College and the School of Mines," 10, no. 2 (October 1914), 7–20, 25–26; "Consolidation and Women's Suffrage," 10, no. 3 (November 1914), 24; "Montana State Board . . . Acts," 9, no. 1 (September 1913), 26–27; and "Montana University and Agricultural College Presidencies are Vacant," 10, no. 10 (June 1915), 5. See also William G. Ferguson, Manager, to Morton J. Elrod, Editor, 21 July 1914, MJE, S2, B1, F11.

119. Morton J. Elrod to Members of the Horticultural Society, 22 October 1914; Form Letter, 8 January 1913; and to M. B. Hampton, 29 October 1914; notes from M. B. Hampton, George Armitage, and P. N. Bernard on public meeting, all MJE, S5, B26, F16.

120. For agreement, see Ernest O. Melby, "Organizing Montana's System of Higher Education" (no date, but 1944); and Ernest O. Melby to State Board of Education, 31 March 1945, both RG1, PO, S19, B52, F "Duplication—Melby Attempts to Change Unit Functions, 1944–45"; and Karlin, "Conflict and Crisis," 48–61.

121. Charles Helmick to Morton J. Elrod, 8 January 1913, MJE, S5, B26, F16.

122. Morton J. Elrod, handwritten on consolidation (no date but late 1914), MJE, S5, B26, F19.

Chapter 3

1. Morton J. Elrod, "Montana's New National Park Is Grand, A Priceless Pleasure Ground For All," *Missoulian* (24 July 1910), 1, MJE, S4, B18, F16.

2. Malone and Roeder, *Montana*, ch. 11.

3. Clemens P. Work, *Darkest Before Dawn: Sedition and Free Speech in the American West* (Albuquerque: University of New Mexico Press, 2005), passim; Gutfeld, *Montana's Agony*, passim; and Astrid Honoria Sisson, "Edward Octavius Sisson, May 24, 1869–January 24, 1949" (15 June 1950), passim, RG1, PO, S2, B169, F "President–Sisson, Edward O. (1917–21)."

4. Gutfeld, *Montana's Agony*, ch. 4–6.

5. Clapp, "Narrative," ch. 6, 6*ff.* and 90–100; and "Proposal for Adjustment of Work For Students Registered in the Discontinued German Classes," effective 23 April 1918, "in accordance with the order to the State Defense Council"; and various letters, including Owen Nelson, President of the University of Wyoming, to President E. O. Sisson, 31 May 1918, arguing against suspending German language instruction, "but if public sentiment demands it, we might need to yield"; F. C. Scheuch, Chairman of the Special Committee, to President E. O. Sisson (no date but May–June 1918), reporting arrangements for students

affected by the discontinuance, all RG1, PO, S5, B6, F "CURR. Elimination of German—World War I." Also Merriam, *History,* ch. 5; Board of Education minutes during the War years, esp. SBE, Minutes, Regular Meeting, June 1918, 74, M418, R2; Arnon Gutfield, "Levine Affair: A Case Study in Academic Freedom," *Pacific Historical Review* 39, no. 1 (February 1970), 25–26; and Stearns, "Fisher," esp. 11–25.

6. Ira Katznelson, *Fear Itself: The New Deal and the Origins of Our Time* (New York: Liveright Publishing Company, 2013), passim.

7. See unknown (perhaps Sisson or Scheuch), "German" (no date but early 1918, as the text mentioned an article in the *Modern Language Journal* 2, no. 2 (February 1918), RG1, PO, S5, B6, F "CURR. Elimination of German—World War I."

8. Merriam, *History,* 50–52; Sisson, "Edward Octavius Sisson," passim, RG1, PO, S2, B169, F "President–Sisson, Edward O. (1917–21)"; and President Edward O. Sisson, "An Open Letter Concerning the State University of Montana," reprint from *Missoulian* (17 January 1921), RG1, PO, S19, B169, F "President—Letters."

9. On the SATC, see Merriam, *History,* 52; and Giltner, "Montana State University," 15 February 1939, 6, RG1, PO, S15, B26, F "ARCHIVES: Informational." Documents concerning the SATC fill a large box of folders, RG1, PO, S3, B18; and Edward O. Sisson, Report to U.S. Bureau of Education, Department of the Interior, 1919, RG1, PO, S19, B46, F "Bureau of Education Reports and Letters—1916–1931." See also "University of Montana Men Who Died during the World War or as a Result of the War," 28 May 1930; and "Roll of Honor, World War I," 27 June 1933, and other documents, all RG1, PO, S15, B26, F "Roll of Honor—World War I."

10. "M. J. Elrod" (no date but 1950s), MJE, SI, B1, F1-1; also Business Manager to Mrs. Lawrence Maloney, 8 December 1925; and M. J. Elrod to A. L. Stone, 28 June 1926, both RG1, PO, S3, B18, F "President's Office, 1921, No. 1."

11. Edward C. Elliott, to President Clapp, 21 April 1922; J. E. Kirkwood, member of the committee on campus development, to President C. H. Clapp, 17 April 1922, with renderings of the "Soldiers Memorial . . . for the center of the campus oval," all RG1, PO S6, B42, F "Campus Development, 1922–44."

12. Gutfeld, *Montana's Agony,* ch. 2–3 and 6–7.

13. "Palmer Raids," at http://en.wikipedia.org/wiki/Palmer_raids.

14. James H. Bonner, Chairman, Charles E. Mallett, William M. Aber, M. J. Elrod, and J. H. Underwood, physical plant committee, to President F. C. Scheuch, 14 October 1916, RG1, PO, S4, B63, F "General to 1933"; SBE, Minutes, Regular Meeting, 3 April 1920, 342, M418, R2; Merriam, *History,* 64–66; "Cass Gilbert" at http://en.wikipedia.org/wiki/Cass_Gilbert; and "The Gilbert–Carsley Plan" at http://content.lib.umt.edu/omeka/items/show/864; also "A History of Campus Planning," at http://content.lib.umt.edu/omeka/exhibits/show/a-history-of-campus-planning/1893-1929/3; and anonymous

(J. B. Speer), "Montana State University, Missoula, Proposed Land Acquisitions For Campus Purposes Requiring Special Legislative Appropriations," 1 January 1943, with map, RG1, PO, S3, B16, F "1945 Legislative Data—MSU."

15. W. F. Brewer, "If it's for the Schools, Everybody's for it" (no date but 1920), campaign pamphlet with copies of the initiatives, RG1, PO, S4, B168 (file identifier missing); and various materials, RG1, PO, S4, B167, F "Funds Campaign, 1920—Advertisements."

16. Edward C. Elliott, "The Future of the University: Charter Day—State University," 18 February 1921; and Edward C. Elliott to Dean A. L. Stone, 7 March 1921, both RG1, PO, S4, B167, F "Charter Day—Addresses, 1912, 1914, 1918, 1919, 1920, 1921."

17. The Phi Beta Kappa Association of the State University of Montana, "Essential Facts For Consideration In Connection With Its Petition For a Charter In Phi Beta Kappa" (no date but 1929), RG1, PO, S15, B26, F "ARCHIVES: Informational"; and Merriam, *History*, 70.

18. Morton J. Elrod, "Charter Day Address," 1916, MJE, S5, B26, F13.

19. Morton J. Elrod, *College, Past and Present* (Bloomington, IL: University Press, 1899), esp. 23–26.

20. Harry R. Lewis, *Excellence Without a Soul: How a Great University Forgot Education* (New York: Public Affairs, 2006), passim. Also Louis Menand, *Marketplace of Ideas* (New York and London: W. W. Norton and Company, 2010), ch. 1–3; Earl J. McGrath, *Graduate School and the Decline of Liberal Education* (New York: Institute of Higher Education, 1959), passim; and Earl J. McGrath and Charles H. Russell, *Are Liberal Arts Colleges Becoming Professional Schools* (New York: Institute of Higher Education, 1958), passim.

21. M. A. Brannon to M. J. Elrod, 1 November 1923, RG30, HR, F "Elrod, M. J."; and SBE, Minutes, Regular Meeting, June 1918, 134, M418, R2. This change assigned an administrator to do work the faculty had previously done, a change that attained maturity after World War II, as see Speer, "A Bird's-Eye View," *Journal of Higher Education* 2, no. 9 (December 1933), 461–67, RG1, PO, S19, B46, F "Questionnaires."

22. See unsigned letter to Morton J. Elrod, 26 June 1925; unsigned letter to Floyd W. Gail (and others), 20 November 1925, Elrod as Chair, MJE, S3, B5, F19. Also Morton J. Elrod, "To Science Teachers of the Northwest: The Results of the second part of the study," 12 October 1927, MJE, S3, B5, F2; Morton J. Elrod, "Report of the Committee on Science Courses," 8 April 1930, MJE, S3, B5, F22; Melvin A. Brannon to Morton J. Elrod, 11 April 1928, MJE, S3, B5, F23; and Morton J. Elrod, "To the Members of the Committee for General Science, Inland Empire Science Teachers Association" (no date, but 1929 or 1930), MJE, S3, B5, F27. For essential agreement with Elrod, see Carl A. Jessen, Montana High School Supervisor, to Morton J. Elrod, 29 September 1927, enclosing "Parts I, II, and III of the tentative science course of study for Montana high schools," MJE, S3, B6, F6.

23. Morton J. Elrod, "The Heritage of Youth," 1928, with photographs, MJE, S3, B5, F16.

24. Morton J. Elrod, "The American University" (no date but after 1916), MJE, S3, B4, F9. See also Elrod, *College, Past and Present* (1899), passim, for differences of emphasis; in 1899, he held a more optimistic view of the group approach to requirements and electives. For a modern but similar critique, Richard F. O'Donnell, "Higher Education's Faculty Productivity Gap: The Cost to Students, Parents and Taxpayers," 2011, originally available at www.highered ideas.com, but retrieved at http://gaia.pge.utexas.edu/papers/Higher%20Eds %20Faculty%20Productivity%20Gap.pdf. On externally administered admission tests, which Elrod advocated, see Rebecca Zwick, "College Admission Testing," Arlington, VA: National Association for College Admission Counseling, 2007, at www.nacacnet.org/media-center/standardizedtesting/documents /standardizedtestingwhitepaper.pdf.

25. For a similar modern development, see Donna M. Desrochers and Rita Kirshstein, "Labor Intensive or Labor Expensive? Changing Staffing and Compensation Patterns in Higher Education" (Delta Cost Project, American Institutes for Research, February 2014), radical increases in part-time, temporary faculty members, at http://www.air.org/sites/default/files/downloads /report/DeltaCostAIR-Labor-Expensive-Higher-Education-Staffing-Brief-Feb 2014.pdf.

26. For modern agreement, see Menand, *Marketplace of Ideas,* ch. 1–2.

27. Except as specifically noted, the following discussion draws on and quotes from Elrod, "Heritage of Youth," 1928, MJE, S3, B5, F16. For an apt description of Elrod's view of history as progress through social acquisition rather than biological evolution, see Edward Hallett Carr, *What is History?* (New York: Vantage Books, 1961), ch. 5. Also Wootton, *Invention of Science,* 345–48.

28. Morton J. Elrod, "Raising the Level," *Missoulian* (6 December 1920), 1, 5, originally a Cosmos Club paper, MJE, S8, B37, F3; and Morton J. Elrod to Emma, 5 December 1920, MJE, S2, B2, F3. Also Leonard, *Illiberal Reformers,* ch. 6–7.

29. For discussion, see Segerstralle, *Nature's Oracle,* ch. 5, esp. 73–75.

30. Elrod, "The College, Past and Present," 1–10, 22–26.

31. Elrod, "Department of Biology," in Craig, President's Report, 1903, 40–45, UniPub, S2, B1895–1912; and Elrod to Furst, 1916, emphasis supplied, MJE, S2, B2, F2.

32. "Joseph Collins (neurologist), 1866–1950, " at http://en.wikipedia.org /wiki/Joseph_Collins_(neurologist).

33. "Edward C. Elliott," at http://en.wikipedia.org/wiki/Edward_C._Elliott.

34. SBE, Minutes, Special Meeting, 11 October 1915, 93–95, on Elliott's conditions for the appointment, M418, R2. See also Elrod, "Montana State Board . . . Acts," *Inter-Mountain Educator* (September 1913), 26–27.

35. For background, Hamilton, *Academic Ethics,* passim. Also SBE, Minutes, Special Meeting, 22 January 1917, 173, on a "gag rule" by Elliott, M418, R2; and "University Act No. 514, Chancellor's Calendar–E, December 1917," RG1, PO, S2, B25, F "Investigations and Censure Prior to 1936."

36. Merriam, *History,* 64, quoting Clapp to Chancellor-elect Melvin A. Brannon, 20 December 1922.

37. Clapp's comments, Minutes, Committee on Budget and University Policy, 10 October 1922, RG1, PO, S6, B42, F "Budget and Policy to 1936"; also Charles H. Clapp, "Remarks on Centralized Systems, National Association of State Universities, November 1933"; and C. H. Clapp to President A. H. Upham, Miami, Ohio, both RG1, PO, S5, B6, F "National Association of State Universities, 1910–1951."

38. Morton J. Elrod to Chancellor Edward C. Elliott, 13, 14, and 15 January 1916, MJE, S2, B2, F1.

39. SBE, Minutes, Regular Meeting, June 1918, 148–49, M418, R2, on the powers and duties of the chancellor.

40. SBE, Minutes, Special Meeting, 11 October 1915, 93–95; and SBE, Minutes, Regular Meeting, June 1918, 59–154, at 145–47, M418, R2.

41. SBE, Minutes, Regular Meeting, 7 June 1915, 79–89; SBE, Minutes, Special Meeting, 11 October 1915, 95–97; and SBE, Minutes, Special Meeting, 28 April 1916, 114–15, M418, R2; Kofoid, "Academic Freedom and Academic Tenure," *Bulletin of the American Association of University Professors* 3, 28–42; "Through Personal Revenge Professors Are Dismissed," *Missoulian* (9 July 1915), MJE, S5, B26, F17; Karlin, "Conflict and Crisis," 48–61; Merriam, *History,* 24, footnote 10, and 42; Clapp, "Narrative," ch. 5, 8–10; and anonymous (probably F. C. Scheuch,) to Chancellor Edward C. Elliott, 8 November 1915, RG1, PO, S7, B29, F "RYMAN, J. H. T."

42. Quoting Elliott, see Kofoid, "Academic Freedom and Academic Tenure," *Bulletin of the American Association of University Professors* 3, 41.

43. For appointments, tenure of office, and other policies placing the chancellor in charge, see Edward C. Elliott, "Administrative Memorandum No. 180, Relating To: The University Code," 20 October 1921, 66 (typescript), RG1, PO, S4, B167, F "Univ Code (First Draft of Part I [actually Part II]"; and Office of the Chancellor, "Administrative Memorandum No. 100," 29 June 1918, "Regulations in Regard to Tenure . . . ," University Act No. 673 (22 June 1918), RG1, PO, S5, B47, F "Salaries, 1918–1944"; SBE, Minutes, Special Meeting, 22 January 1917, 173, a gag rule by Elliott, M418, R2; and "University Act No. 514, Chancellor's Calendar–E, December 1917," RG1, PO, S2, B25, F "Investigations and Censure Prior to 1936."

44. Morton J. Elrod, handwritten draft of the Committee Report on the Louis Levine case (no date but April 1919), MJE, S5, B28, F3. Also E. O. Sisson to the Service Committee, Professor Elrod, Chairman, 20 May 1919; and Morton J. Elrod, Chairman, Paul C. Phillips, and Walter L. Pope, Committee

on Service, to President E. O. Sisson, 3 July 1919, both RG1, PO, S5, B47, F "Tenure."

45. Clapp, "Narrative," ch. 6, 30–31. Clapp confused the state university welfare committee with the mandated committee on service.

46. See unknown (undoubtedly Sisson), "Welfare Committee," handwritten title, one typed sheet with the date of "Feb. 1918"; and excerpts "From the Faculty Minutes of January 15, 1918: Welfare Committee"; "From the Faculty Minutes, November 26, 1918"; "From the Faculty Minutes, March 11, 1919"; E. O. Sisson, President, to The Welfare Committee, 20 March 1918; and E. O. Sisson to J. J. Thornleer, Secretary to the Faculty, University of Arizona, 13 February 1918, all RG1, PO, S5, B23, F "Welfare Committee, 1918."

47. See various materials related to the Louis Levine case, including Elrod's handwritten draft of the committee report, MJE, S5, B28, F3; Morton J. Elrod, Chairman, Paul C. Phillips, and Walter J. Pope, "Findings of the Committee on Service . . . ," 1 April 1919, 6, RG30, HR, F "Levine, Louis"; Morton J. Elrod, Chairman, "Findings of the Committee on Service of the Montana State University In the Case of Arthur Fisher," 13 September 1921, MJE, S5, B27, F7; and Morton J. Elrod, "The Committee on Service Report," various years, in Charles H. Clapp, President's Report, 1921–34, UniPub, S2, B1912–1915, B1921–1929, and B1929–1934. For Elrod's service on AAUP subcommittees investigating complaints, see F. S. Deibler, Northwestern University, to Morton J. Elrod, 30 October 1920, the Middlebury College case, MJE, S2, B2, F5; Philo F. Hammond to A. G. Crane, 22 December 1926, MJE, S2, B2, F11; and John M. Magnina, Legal Advisor to Committee A, to Morton J. Elrod, 4 August 1928, MJE, S2, B3, F1, the last two on the Gossard case in Wyoming.

48. Elrod, "Findings of the Committee on Service," 1 April 1919; and F. S. Deibler to President E. O. Sisson, 23 March 1919, both RG30, HR, F "Levine, Louis"; also Gutfeld, "The Levine Affair," 19–37; Gutfeld, *Montana's Agony*, ch. 10; and Louis Levine, *Taxation of Mines in Montana* (Reprint ed.; Boulder: University of Colorado BiblioLife, n.d.), passim, originally published in 1919.

49. For alleged company pressure, see Gutfeld, *Montana's Agony*, 106–108; Merriam, *History*, 54; and Howard, *Montana: High, Wide, and Handsome*, ch. 23, "A Russian Jew Named Levine. . . . "

50. Gutfeld, "The Levine Affair," 19–37; Gutfeld, *Montana's Agony*, ch. 10 and 12; Malone and Roeder, *Montana*, 219–22.

51. See Elrod's handwritten draft of the committee report, MJE, S5, B28, F3. Also "Abbott Lawrence Lowell," at http://en.wikipedia.org/wiki/Abbott _Lawrence_Lowell.

52. Morton J. Elrod, "Montana Professor Suspended," and "Dr. Louis Levine Reinstated," *Inter-Mountain Educator* 14, no. 6 (February 1919), 24–26; and 14, no. 8 (April 1919), 41–42.

53. SBE, Minutes, Regular Meeting, 7 April 1919, 250, M418, R2.

54. Clapp, "Narrative," ch. 6, 23–36.

55. Astrid Honoria Sisson, "Edward Octavius Sisson, May 24 1869–January 24, 1949" (15 June 1950), ch. 4, RG1, President's Office, S2, B169, F "President Sisson, Edward O. (1917–21)."

56. Elrod, "Dr. Louis Levine Reinstated," *Inter-Mountain Educator,* 41–42.

57. F. S. Deibler to President E. O. Sisson, 23 March 1919, RG30, HR, F "Levine, Louis."

58. Edward O. Sisson to Chancellor Edward C. Elliott, 29 September 1919; Louis Levine to President E. O. Sisson, 1 October 1919; E. C. Elliott to E. O. Sisson, 10 October 1919; and E. O. Sisson to Louis Levine, 13 October 1919, all RG30, HR, F "Levine, Louis."

59. Frederick S. Deibler, "Committee on Academic Freedom: Statement on the Case of Professor Louis Levine of the University of Montana," AAUP, *Bulletin of the American Association of University Professors. Reports of Actions on Carnegie Foundation Proposals; University of Montana; Bethany College; List of Committees* (Boston: AAUP March 1919), 13–25, restating the committee report, since the Board decision mooted the case.

60. Elrod, "Findings of the Committee on Service of the Montana State University In the Case of Arthur Fisher," 13 September 1921, MJE, S5, B27, F7.

61. Stearns, "Fisher," passim.

62. Clapp, "Narrative," ch. 6, 6*ff.* and 90–100; Merriam, *History,* ch. 5; and Stearns, "Fisher," esp. 11–25. For background, Gutfeld, *Montana's Agony,* esp. ch. 11; and Work, *Darkest Before Dawn,* passim.

63. Elrod, "Findings of the Committee on Service of the Montana State University In the Case of Arthur Fisher," 13 September 1921, MJE, S5, B27, F7. See also Edward C. Elliott, "Memorandum re American Legion Charges Against Arthur Fisher," 1921; A. Atkinson, President, Montana State College, to President C. H. Clapp, State University, 10 September 1921, on Fisher's attitude toward his university responsibility; and C. W. Leaphart and C. H. Clapp to Montana State Board of Education, 16 September 1921, all RG30, HR, F "Arthur Fisher"; also various documents in a separate file, RG30, HR, F "Staff: Fisher, Arthur." C. W. Leaphart to the Service Committee, 10 September 1921, MJE, S5, B27, F5; and Charles H. Clapp to A. W. Vernon of the AAUP, 5 July 1923, as quoted in Stearns, "Fisher," 99. Gutfeld, *Montana's Agony,* ch. 11, failed to take note of the strong defense of academic freedom by the chancellor, president, and dean, or the fact that Fisher had only a term contract.

64. SBE, Minutes, Regular Meeting, 17 September 1921, 103–16, at 111–12, M418, R2; Stearns, "Fisher," 90–96; "Prof. Fisher Dismissed From University," *Missoulian* (19 September 1921), despite the headline error; and "The Case of Arthur Fisher," *Missoulian* (20 September 1921) MJE, S5, B7, F10.

65. F. S. Deibler, "Report on the University of Montana," *Bulletin of the American Association of University Professors* 10, no. 3 (March 1924), 154–62,

not technically a decision since the lapsing of Fisher's term contract in 1922 rendered the case moot.

66. Morton J. Elrod, "Professor Fisher," *Inter-Mountain Educator* 17, no. 2 (October 1921), 73–74.

67. Elliott, "Memorandum re American Legion Charges," 1921, RG30, HR, F "Arthur Fisher."

68. Morton J. Elrod, "Play Fair, Boys," *Inter-Mountain Educator* 17, no. 5 (January 1922), 216.

69. Elrod, "The Inter-Mountain Educator," *Montana Educator* 1, no. 3 (November 1924), 15–16.

70. SBE, Minutes, Regular Meeting, 19 December 1921, 127, M418, R2.

71. M. J. Elrod to President Clapp, 4 November 1921; and President Clapp to M. J. Elrod, 4 November 1921, both RG1, PO, S6, B82, F "Service Committee."

72. Merriam, *History*, 66; and Fred Wyman to G. M. Sheridan, Sigma Nu, 8 December 1932, urging two students to rescind their resignations from the *Sentinel* (which they submitted because Elrod and Dean Stone refused to approve the students' wording), with a note to Elrod hoping "this letter meets with your approval," MJE, S2, B3, F5. But for Merriam's passionate and successful response to President Clapp's stated intent to prevent distribution of the *Frontier*—a journal Merriam owned as an outlet for creative writing students and young writers—because of word usage in some of the student articles, see H. G. Merriam to President C. H. Clapp, 22 November 1922, RG1, PO, S19, B46, F "Frontier."

73. Elrod, "Findings of the Committee on Service of the Montana State University In the Case of Arthur Fisher," 13 September 1921, MJE, S5, B27, F7; and C. H. Clapp, President, to Professor John Hollen, University of Texas, 13 February 1930, responding to Hollen's queries on academic freedom and outside activities of faculty, RG1, PO, S5, B47, F "Tenure."

74. SBE, Minutes, Special Meeting, 12 December 1919, 331–32; SBE, Minutes, Regular Meeting, 12 July 1920, 364; and SBE, Minutes, Regular Meeting, 6 December 1920, 1–4, M418, R2.

75. On Merriam, see "Guide to H. G. Merriam Papers, 1890–1980," at http://nwda.orbiscascade.org/ark:/80444/xv78663; and Merriam, *History*, passim. For the "Memorial" of 16 March 1921, see H. G. Merriam to CHC, 18 February 1933, and Edward C. Elliott, Chancellor, to President C. H. Clapp, 16 July 1921, RG1, PO, S5, B6, F "CURR. Curriculum Revision, 1933," and F "CURR. Curriculum General to 1942," respectively.

76. Merriam, *History*, 56–57; G. B. Castle, Chair, Montana State University Budget and Policy Committee, 12 February 1945, attached to E. G. Marble, "Minutes," Faculty Meeting, 25 January 1945, RG1, PO, S6, B42, F "Faculty Minutes, 1940–Aug. 1946"; and "Excerpts from Faculty Minutes Pertaining to Budget and Policy Committee," 12 April 1921 through 11 March 1936, RG1, PO, S6, B42, F "Budget and Policy to 1936."

77. Merriam, *History,* 56–59. See also Speer, "A Bird's-Eye View," *Journal of Higher Education* 2, no. 9 (December 1933), 461–67, RG1, PO, S19, B46, F "Questionnaires," for the extraordinary influence of the committee.

78. As quoted in Merriam, *History,* 69.

79. President C. H. Clapp to President Alfred Atkinson, 10 March 1926, RG1, PO, S6, B42, F "Budget and Policy to 1936."

80. Budget and Policy Committee, Minutes, 10 October 1922, unanimous approval in 1922 to authorize "one man committees" and midlevel administrators to handle routine administrative work, thus relieving the faculty, RG1, PO, S6, B42, F "Budget and Policy to 1936." For a description of administrative structure at the time, see Speer, "A Bird's-Eye View," *Journal of Higher Education* 2, no. 9 (December 1933), 461–67, RG1, PO, S19, B46, F "Questionnaires."

81. Clapp to Atkinson, 10 March 1926, RG1, PO, S6, B42, F "Budget and Policy to 1936."

82. See various documents, specifically "Excerpts from Faculty Minutes Pertaining to Budget and Policy Committee," 12 April 1921 through 11 March 1936, with annual reports and minutes during that period, RG1, PO, S6, B42, F "Budget and Policy to 1936."

83. Merriam to CHC, 18 February 1933, the "Memorial" attached to a handwritten note urging another such "Memorial" in 1933; and Edward C. Elliott, Chancellor, to President C. H. Clapp, 16 July 1921, the "Memorial" attached, RG1, PO, S5, B6, F "CURR. Curriculum Revision, 1933" and F "CURR. Curriculum General to 1942," respectively; Merriam, *History,* 56–59; Edward C. Elliott, Chancellor, to President C. H. Clapp, 16 July 1921; J. H. Underwood to President Clapp, 21 October 1921; H. G. Merriam to President C. H. Clapp, 10 December 1921; Melvin A. Brennan, Chancellor, to President C. H. Clapp, 9 December 1924; and President C. H. Clapp to Chancellor M. A. Brannon, 12 December 1924, all RG1, PO, S5, B6, F "CURR. Curriculum General to 1942."

84. C. H. Clapp, President, to Curriculum Committee, 28 November 1921, RG1, PO, S5, B6, F "CURR. Curriculum General to 1942."

85. See examples of reform at other institutions, such as the University of Chicago, and various memoranda, letters, and other materials, RG1, PO, S7, B6, F "Curriculum Revision, 1933." For discussion of curricular reform, see Lewis, *Excellence Without a Soul,* passim, esp. 305.

86. Curriculum Committee Minutes, 9 December 1930, RG1, PO, S7, B14, F "Minutes, Curriculum Committee, 1928–1954." For a participant's remembrances, see Merriam, *History,* 60–82.

87. Charles H. Clapp, President's Report, 1933, 4, UniPub, S2, B1929–1934; Charles H. Clapp, President's Report, 1934, 4, UniPub, S2, B1929–1934; Clapp, "Narrative," ch. 7, 32*ff.*; and Lewis, *Excellence Without a Soul,* passim.

88. Merriam, *History,* ch. 8–10, for a general discussion. A definitive history of these changes does not exist.

89. Morton J. Elrod, "Division of Biological Sciences," 19, in Clapp, President's Report, 1934, UniPub, S2, B1929–1934; and Morton J. Elrod, "Introduction to Biological Science, 1933," RG1, PO, S8, B6, F "Survey Courses."

Chapter 4

1. Morton J. Elrod, "Introduction," handwritten (no date but probably 1926–30), for a planned history of the park, plus covering his first trips in 1906, 1909, 1910, 1911, and 1914, "Early Expeditions to Glacier National Park, 1906–1914," MJE, S4, B15, F12; and Morton J, Elrod, handwritten notes on the park (no dates but probably 1914–28), MJE, S4, B18, F7.

2. Morton J. Elrod to Emma, 2 July 1901, MJE, S2, B1, F9.

3. Elrod, "Introduction" (1926–30), MJE, S4, B15, F7 and F12; and notes on the park (1914–28), MJE, S4, B18, F7.

4. Morton J. Elrod, numerous clippings, including one describing a trip to the Canadian boundary, *Missoulian* (5 September 1910), 1, 5, MJE, S4, B14, F21.

5. Elrod, "Introduction" (1926–30), MJE, S4, B15, F7 and F12; also Morton J. Elrod, "The Biological Station," 45–46, in Craig, President's Report, 1906, S2, B1895–1912; and Morton J. Elrod, "The University of Montana Biological Station," 1910, MJE, S5, B28, F25. See as well Morton J. Elrod, "A Study of Timber," 1906, MJE, S4, B15, F5; also Morton J. Elrod, "Trip to Sperry Glacier" (no date but 1931); and Morton J. Elrod, "Sperry (Comeau) Pass" (no date but 1931), both MJE, S4, B20, F1; also Morton J. Elrod, fragment on 1906 trip, MJE, S4, B19, F7; and Morton J. Elrod, "The Attempt on St. Nicholas" (no date but 1906), MJE, S4, B19, F9–11. The introduction dated the first excursion in 1908, but the Biological Station report for 1906 had the details. Elrod, "Proposed Glacier National Park" (no date but written before May 1910), MJE, S4, B13, F15.

6. Elrod, "Introduction" (1926–30), MJE, S4, B15, F7 and F12. Also see Morton J. Elrod, "Blackfoot Glacier in Montana's National Park: Interesting Feature of that Pleasure Ground," *Missoulian* (16 July 1911), MJE, S4, B14, F22; and Morton J. Elrod, typescript, "Blackfoot Glacier," 1911, MJE, S4, B15, F4. See also Morton J. Elrod, "Gunsight Pass," 1911, MJE, S4, B18, F8; and Elrod, "Proposed Glacier National Park" (no date but written before May 1910), MJE, S4, B13, F15.

7. Elrod, "Proposed Glacier National Park" (no date but written before May 1910), MJE, S4, B13, F15.

8. C. W. Buchholz, *Man in Glacier,* ch. 1–5, at http://www.nps.gov/history /history/online_books/glac2/. See also Harper, "Conceiving Nature," 3–24, 91–94; Christopher S. Ashby, "Blackfeet Agreement of 1895 and Glacier National Park: A case history" (1985), ch. 3, at http://scholarworks.umt.edu/cgi/view content.cgi?article=2703&context=etd; and the chronicle of prospecting on

the Blackfeet Reservation land and the timber preserve leading to open public access in 1898, George Bird Grinnell to R. S. Yard, 27 October 1927, and Robert Sterling Yard, "The Copper Rush," to Mr. Grinnell, 25 September 1917, both GBG, R42.

9. Elrod, "Montana's New National Park is Grand: A Priceless Pleasure Ground for All," *Missoulian* (21 July 1910), with photographs of Gunsight Lake, Lake Louise, Sperry Glacier, and Blackfoot Glacier, taken in 1906 and 1909; also see A. L. Stone, "Glacier Park," *Missoulian* (21 July 1910), both R39. For his advocacy prior to the establishment of the park, see Elrod, "Proposed Glacier National Park" (written before 11 May 1910), MJE, S4, B18, F15. See also Ashby, "Blackfeet Agreement," ch. 4.

10. Diettert, *Grinnell's Glacier,* ch. 6–9; and "George Bird Grinnell," at http://en.wikipedia.org/wiki/George_Bird_Grinnell. For some three decades, Grinnell edited *Forest and Stream,* the voice of conservation in the country. For Grinnell's long-time interest in and effort for the park legislation, see George Bird Grinnell to Dr. H. C. Bumpus, 4 May 1910, GBG, R14, where he wrote "a 'baby' of mine for pretty nearly twenty-five years."

11. Ashby, "Blackfeet Agreement," ch. 4; and Larry Len Peterson, *Call of the Mountains: The Artists of Glacier National Park* (Tucson, AZ: Settlers West Galleries, 2002), 7, 2–11.

12. Morton J. Elrod, fragment on the park legislation (no date but probably written in 1909–1910), MJE, S4, B19, F7. See the Dixon Papers, JMD, MSS 055.

13. Ashby, "Blackfeet Agreement," 38–40, citing several Senate and House discussions and reports, with some misstatements but accurate as to timing and detail of the bill.

14. The amendments related to identifiable boundaries, railroad access, water development in the park, protection of pre-existing property rights, and prevention of any transfers of possible railroad grants in the park area. For these details, see *Congressional Record—Senate* and *Congressional Record—House* (11 December 1907, 24 February 1908, 29 April 1908, 1 May 1908, 20 January 1910, 25 January 1910, 9 February 1910, 15 April 1910, 2–3 May 1910, and 11 May 1910), vol. 42, part 1, 269; part 3, 2366; and part 6, 5319 and 5514; and vol. 45, part 1, 832 and 958–61; part 2, 1639–41; and part 5, 4640–41, 4669, 5342, 5431, 5570, and 5691; and part 6, 5697 and 6322. Also "Senate Report No. 580," *Senate Reports,* 60th Congress, 1st Session, vol. 2, series vol. 5219; "Senate Report No. 106," *Senate Reports,* 61st Congress, 2nd Session, vol. 1, series vol. 5582; and "House Report No. 767," *House Reports,* 61st Congress, 2nd Session, series vol. 5592.

15. For passage of the Act and signature by President William Howard Taft on 11 May 1910, with no mention of Elrod, see Harper, "Conceiving Nature," 3–24, 91–94; Diettert, *Grinnell's Glacier,* ch. 7–8; and various Grinnell letters, especially George Bird Grinnell to A. Chamberlain, 4 May 1910, GBG, R14; and

"Calendar No. 52," Senate, Glacier Park Bill Report of Senator Joseph M. Dixon, 29 April 1908, GBG, R42. See also W. T. Hornaday to Senator Joseph M. Dixon, 13 March 1909; and Senator Joseph M. Dixon to Dr. Hornaday, 18 March 1909, JMD, MSS 055, B10, F6; and Buchholz, *Man in Glacier*, ch. 4, at http://www.nps .gov/history/history/online_books/glac2/.

16. Karlin, *Dixon of Montana* 1, ch. 6–9; and Harper, "Creating Nature," esp. 21.

17. Morton J. Elrod, "Congress Passes Bill Creating Glacier National Park" (no date, but probably after 1927 based on internal references), MJE, S4, B18, F14, handwritten and typed copies, probably as a chapter for the park history he intended to write, including several of his own photographs, other photographs, various reports, letters from the secretaries of the interior and agriculture, and a detailed topographical description and map of the park prepared by R. H. Chapman, who subsequently served as acting supervisor; and Elrod, "Proposed Glacier National Park," with "Photographs by the Author" (written prior to passage of the act in 1910, in a folder marked 1907), MJE, S4, B18, F15, fifty-eight handwritten pages, including descriptions of his trips to the park area in 1906 and 1909.

18. Buchholz, *Man in Glacier*, ch. 5, at http://www.nps.gov/history/history /online_books/glac2/. See Ashby, "Blackfeet Agreement," ch. 4; Harper, "Creating Nature," passim; Peterson, *Call of the Mountains*, ch. 2, for the support provided by the Hill family. On the Boone and Crocket Club and Grinnell's founding role and promotion of preserving the St. Mary region in Montana, see Diettert, *Grinnell's Glacier*, ch. 5. On the park's function, see Elrod, "Proposed Glacier National Park" (written before 11 May 1910), MJE, S4, B18, F15; and Elrod, "Montana's New National Park," *Missoulian* (21 July 1910), R39.

19. Harper, "Conceiving Nature," passim.

20. Elrod, "Proposed Glacier National Park" (written before 11 May 1910), MJE, S4, B18, F15.

21. Supporters developed this same theme in the Congressional debates, see *Congressional Record*, part 1, 959–961 and part 2, 1639–1641, in the Senate; and part 5, 4640–4641, 4669, 5342, 5431, 5570, 5697, in the House.

22. Elrod, "Montana's New National Park is Grand: A Priceless Pleasure Ground for All," *Missoulian* (21 July 1910), R39.

23. Elrod, "Introduction" (1926–30), MJE, S4, B15, F7 and F12. See also Walter Lehman, Lewistown, to Morton J. Elrod, 29 December 1950, MJE, S2, B3, F11; Elrod, "The Biological Station," 53–55, in Duniway, President's Report, 1909–1910, S2, B1895–1912; Morton J. Elrod, "The Garden Wall in Glacier National Park Presents a Scene of Great Beauty Unexcelled," *Missoulian* (25 June 1911), 3, MJE, S4, B15, F21; and Morton J. Elrod, "M. J. Elrod Returns From Glacier Park," *Missoulian* (5 September 1910), 1, 5, MJE, S4, B14, F21. On "Adair's Place," see "Historic Place Names," in National Park Service, "Glacier: Through the Years in Glacier National Park," at http://www.nps.gov/history/history

/online_books/glac/appa.htm, for Adair's Ridge near Polebridge; and Harper, "Conceiving Nature" (2010), 7, for a photograph of the Adair boarding house.

24. Buchholz, *Man in Glacier*, ch. 5, concerning fire damage, at http://www.nps.gov/history/history/online_books/glac2/, with the largest fire in the Northwest in 1910.

25. Morton J. Elrod, handwritten notes on "First Map of Glacier National Park" (undated, but 1912–13), MJE, S4, B18, F7. For Grinnell's mapping of the park region, see Diettert, *Grinnell's Glacier*, ch. 5. And for Elrod's possession of R. H. Chapman's topographical map containing many names, probably the one he used to prepare the analysis, see Elrod, "Congress Passes Bill Creating Glacier National Park" (no date, but probably 1927), MJE, S4, B18, F14.

26. Morton J. Elrod, "Glacier Park Spots" (no date but 1912 or 1913), MJE, S4, B18, F13; and Morton J. Elrod, fragment on park superintendents and chief rangers (no date, but listing ended in 1921), MJE, S4, B19, F7.

27. Morton J. Elrod, "The Relationship of the People to the Glacier National Park" (no date but 1923), Elrod reversed the wording in the title from "the Glacier National Park to the People" to "the People to the Glacier National Park," MJE, S4, B18, F17.

28. Morton J. Elrod, "Glacier Not Meant As Automobile Park," *Missoulian* (no date but 1926), 1, 12, MJE, S4, B14, F23.

29. Morton J. Elrod, "The Transmountain Road in Glacier Park" (undated but 1925 or 1926); Morton J. Elrod, "Glacier Park's Transmountain Road is a Marvel," *Missoulian* (28 February 1926), 1, 2; Morton J. Elrod, "The Great Glacier Park Road," *Missoulian* (25 November 1928), with photographs, showing arrival at Logan Pass; and Morton J. Elrod, several more accounts, all MJE, S4, B20, F7. Also W. G. Peters, Associate Highway Engineer, Transmountain Highway, to Morton J. Elrod, 7 August 1928, mentioning the Logan Pass accomplishment and Elrod's trip to the pass, offering to share videos and a movie, MJE, S4, B14, F13. See also Morton J. Elrod to Emma, 2 August 1928, for Elrod's planned trip to Logan Pass, MJE, S2, B3, F1.

30. Morton J. Elrod, "Elrod Compares Modes of Travel," *Missoulian* (no date, but 1932–33), MJE, S4, B20, F7; and "Local People Thrill in Mamer-Plane Trip," *Missoulian* (no date, but 1932–33), this trip did not go over the Park because of weather conditions, MJE, S4, B14, F21. On Mamer and the *West Wind*, see "Nicholas Mamer," http://earlyaviators.com/emamer.htm.

31. Elrod, "Relationship of the People" (1923), MJE, S4, B18, F17. During the 1920s, he wrote Emma often about the disgusting jazz music that tourists wanted for listening and dancing, MJE, S2, B2–3, F various.

32. C. Reeker, Chief Clerk, Dept. of Interior, to Morton J. Elrod, 28 March 1912, MJE, S4, B14, F2. Grinnell had already named many of the mountains and other features; see Deittert, *Grinnell's Glacier*, ch. 5.

33. Morton J. Elrod to Secretary of the Interior, 21 April 1912, MJE, S4, B14, F2.

34. Morton J. Elrod, "Glacier Park Names and Their Origins" (no date but late 1920s), MJE, S4, B17, F15. See also L. O. Vought, one of the first rangers, to Morton J. Elrod, 22 October and 3 November 1926, both MJE, S4, B20, F9; and L. O. Vought to Morton J. Elrod, 3 February 1935, MJE, S4, B14, F19. See as well H. C. Hockett, Mountain View, Canada, to Morton J. Elrod, 12 April 1925, on Altyn, Cracker Lake, Cracker Peak, Stark Point, and others, MJE, S4, B14, F9; and Dr. G. C. Ruhle, Park Naturalist, Glacier National Park, to Morton J. Elrod, 13 May 1929, MJE, S4, B14, F14.

35. Morton J. Elrod to Richard G. Badger, 5 July 1917, Elrod promised to send a manuscript of a history of the park "with pictures," MJE, S2, B2, F3. For background, see Warren L. Hanna, *Stars Over Montana: Men Who Made Glacier National Park History* (West Glacier, MT: The Glacier Natural History Association, 1988), passim.

36. Morton J. Elrod, typescript fragment (late 1920s), MJE, S4, B15, F12; Morton J. Elrod, "William T. Hamilton" (no date but 1926–28), MJE, S4, B15, F18. See also "32. William T. Hamilton," in "Missoula History Minutes," at http://fortmissoulamuseum.org/minutes.php.

37. Morton J. Elrod, "Raphael W. Pumpelly," 1926; and Morton J. Elrod, "Lyman B. Sperry" (no date but late 1920s), both MJE, S4, B15, F18. See also "Raphael W. Pumpelly," at http://en.wikipedia.org/wiki/Raphael_Pumpelly.

38. Morton J. Elrod, "Duncan McDonald" (no date, but 1926–28), MJE, S4, B15, F18. See also Morton J. Elrod, "Lake McDonald," in Morton J. Elrod, "The Glaciers of Glacier National Park," handwritten (no date but 1926–29), MJE, S4, B16, F1. See also "Duncan McDonald," at http://www.scribd.com/doc/49098440/Duncan-McDonald. Professor William Farr, senior fellow at the Center for the Rocky Mountain West and professor emeritus of history University of Montana, provided information about the Nez Perce and Iroquois origins of McDonald's mother.

39. Elrod, "Lyman B. Sperry" (no date but late 1920s), MJE, S4, B15, F18; and Morton J. Elrod, "Sperry Glacier," in Morton J. Elrod, "The Glaciers of Glacier National Park" (1926–29), MJE, S4, B16, F1. See also Diettert, *Grinnell's Glacier*, 79, on Sperry; and "Man in Glacier" at http://www.nps.gov/history/history/online_books/glac2/chap4a.html.

40. Morton J. Elrod, "George Bird Grinnell" (no date but 1928–29), handwritten and typescript (57 pages), MJE, S4, B16, F4–5, reporting the 1884 date in his comment about the "starvation period" for Indians; for a report of his early travels in the region, see Grinnell to R. S. Yard, 27 October 1927, GBG, R42. See also "George Bird Grinnell," at http://en.wikipedia.org/wiki/George_Bird_Grinnell; Morton J. Elrod, handwritten notes on Grinnell's eighteen articles in *Forest and Stream*, MJE, S4, B16, F6–7; and Morton J. Elrod, handwritten notes on Grinnell's "The Crown of the Continent," *Century Magazine* (no date but 1901), 660–72, MJE, S4, B16, F9. See also Diettert, *Grinnell's Glacier*, ch. 3, esp. 21; and Richard Vaughn, "To the Ice: George Bird Grinnell's 1887

Ascent of Grinnell Glacier" (2010), at http://www.repository.law.indiana.edu
/facpub/747/; and Harper, "Conceiving Nature," 5. See also Howard, *Montana:
High, Wide, and Handsome,* 155–56.

41. Morton J. Elrod, fragment on James Willard Schultz (no date but mid-
1920s), MJE, S4, B19, F12. See also Ashby, "Blackfeet Agreement," ch. 4, esp.
34–35; and Peterson, *Call of the Mountains,* 12–15.

42. Morton J. Elrod, "Walter McClintock" (no date but 1926–28), MJE, S4,
B15, F18; Peterson, *Call of the Mountains,* 15–18; and Steven L. Grafe, "Montana:
Lanterns on the Prairie—The Blackfeet Photographs of Walter McClintock," at
http://indigenouspeoplesissues.com/index.php?option=com_content&view=
article&id=14332:montana-lanterns-on-the-priarie-the-blackfeet-photographs
-of-walter-mcclintock&catid=&&Itemid=54.

43. Elrod, "Sperry Glacier" (1926–29), in Elrod, "The Glaciers of Glacier
National Park" (1926–29), MJE, S4, B16, F1. On Milo Apgar, see "Village Inn
At Apgar," Glacier Park Inc., at http://www.glacierparkinc.com/_village_inn_at
_apgar.php; on Frank Geduhn, see C. W. Guthrie, *Great Northern Railway and
Glacier National Park* (Helena, MT: Far West, 2004). See also "Glacier National
Park (U.S.)," at http://en.wikipedia.org/wiki/Glacier_National_Park_(U.S.).

44. Elrod, "The Relationship of the People" (1923), MJE, S4, B18, F17.
See also "Glacier National Park (U.S.)," at http://en.wikipedia.org/wiki/Glacier
_National_Park_(U.S.); and on Grinnell's role in negotiating the purchase, see
Harper, "Conceiving Nature," 7–10; and Diettert, *Grinnell's Glacier,* ch. 6–7, for
a less favorable view of Grinnell's negotiations and the establishment of the pre-
serve. See also Ashby, "Blackfeet Agreement," ch. 3–4.

45. For the first quotation, see Elrod, *Elrod's Guide* 119, MJE, S4, B15, F15.
For the other quotations, see Elrod, "The Relationship of the People" (1923),
MJE, S4, B18, F17. For a photograph of a miner on Stark Point, see Harper,
"Conceiving Nature," 23. See Diettert, *Grinnell's Glacier,* ch. 7, on mining and
oil prospecting.

46. Harper, "Conceiving Nature," 6.

47. The secretary of the interior specifically inserted an amendment in
the Act reserving authority for water development in the park; see Harper,
"Conceiving Nature," 19. However, Elrod prevailed in this instance. See Morton J. Elrod, "The Effects on Glacier National Park Should the Proposed Dam
at Lower St. Mary Lake Be Constructed" (no date but early 1920s), MJE, S4,
B15, F10; and Morton J. Elrod, handwritten descriptions of the photographs
used in his argument (no date but early 1920s), MJE, S4, B22, F6; also J. R.
Eakin, Superintendent, Glacier Park, to Mr. Grinnell, 11 February 1922, GBG,
R42, on the dire effects of the proposed project. On the innumerable mountains, 250 lakes, sixty glaciers, flora, fauna, and trails, see Elrod, "Montana's New
National Park," *Missoulian* (21 July 1910), R39; and Morton J. Elrod, "The Gla-
cier National Park," *Encyclopedia Americana* (1919), which also appeared in the
Missoula Sentinel (26 April 1919), MJE, S4, B15, F11.

48. "Biological Station Ground," designated and granted, 1910, MJE, S5, B30, F3. See also Morton J. Elrod, "University Biological Station," 1908, concerning the Bull Island site, RG1, PO, S15, B30, F "Biological Station, 1900–1929."

49. Morton J. Elrod to Acting President Frederick C. Scheuch, 2 August 1915, MJE, S5, B29, F5. See also Morton J. Elrod to Mr. O. E. Thomas, Receiver, U.S. Land Office, Kalispell, 29 June 1916, MJE, S2, B2, F2. Also SBE, Minutes, Regular Meeting, 4 and 22 December 1916, 168; and Special Meeting, 22 January 1917, 175, M418, R2.

50. "Excerpt From The Laws of Montana—27th Session, 1941—Chapter 108," RG1, PO, S15, B30, F "Biological Station, 1930–1944"; and Beard, "Morton J. Elrod," citation to ch. 108, Laws of Montana, 27th Session, 1941, Section 4, 4.

51. Morton J. Elrod, "Report of the . . . Biological Station," 1921, MJE, S5, B30, F11.

52. Letters of 1943–44, Acts of Congress, and board and state action, RG1, PO, S15, B30, F "Biological Station, 1930–1944."

53. Elrod, "Report of . . . Biological Station," 1904, MJE, S5, B28, F22.

54. Morton J. Elrod to Chancellor Edward C. Elliott, Madison, Wisconsin, 11 January 1916, MJE, S2, B2, F1.

55. Morton J. Elrod to W. A. Clark, 20 April 1909; Morton J. Elrod to W. A. Clark (no date but 1909), Professor H. S. Jennings, Johns Hopkins University, to W. A. Clark, 5 April 1909; Professor Henry S. Ward, University of Nebraska, to W. A. Clark, 6 April 1909; and Maurice Ricker, Des Moines, Iowa, to W. A. Clark, 6 March 1909, all MJE, S2, B1, F10.

56. Morton J. Elrod to Emma, 29 July 1910, MJE, S2, B1, F11. See also Elrod, "The University of Montana Biological Station," 1910, MJE, S5, B23, F6; and Morton J. Elrod, "Biological Station," 53–55, in Duniway, President's Report, 1910, UniPub, S2, B1895–1912.

57. Jones to Duniway, 16 December 1908, MJE, S2, B1, F10.

58. Elrod, "The University of Montana Biological Station," 1910, MJE, S5, B23, F6.

59. Ibid.

60. Morton J. Elrod, "Biological Station," 1912, MJE, S5, B29, F1; and see Clyde A. Duniway, "Selections from . . . December 1911," in Clyde A. Duniway, President's Report, 1912, 24–25, UniPub, S2, B1895–1912. For the local executive board's refusal, see "Resolution" (no date but 1911), State University Local Executive Board, RG1, PO, S15, B30, F "Biological Station, 1900–1929."

61. Morton J. Elrod to Executive Committee, University of Montana, 9 September 1912, MJE, S5, B28, F14.

62. Morton J. Elrod, "The University of Montana Biological Station," 1912, MJE, S5, B28, F18. See also Morton J. Elrod to Racine Boat Company, 5 August 1912; and Racine Boat Company to Morton J. Elrod, 6 February 1913, both MJE, S2, B1, F11.

63. Morton J. Elrod, "The Biological Station" (no date but 1913), MJE, S5, B28, F14.

64. Elrod, "The University of Montana Biological Station," 1912, MJE, S5, B28, F18.

65. Morton J. Elrod to President E. B. Craighead, "Report Biological Station, University of Montana, Yellow Bay, Flathead Lake, Season 1914," RG1, PO, S15, B30, F "Biological Station Reports—1908, 1914, 1928–1929."

66. Morton J. Elrod, "Flathead Lake," 1914, MJE, S5, B29, F5.

67. Morton J. Elrod, "Rise and Fall of Flathead Lake," 1916, MJE, S5, B33, F4.

68. See http://en.wikipedia,org/wiki/Tui_chub.

69. Elrod, "Report Biological Station," 1914, RG1, PO, S15, B30, F "Biological Station Reports—1908, 1914, 1928–1929"; and see Morton J. Elrod, "Report, Biological Station," 1914, MJE, S5, B29, F4.

70. Morton J. Elrod to Chancellor Edward C. Elliot, 7 June 1916, MJE, S5, B29, F9.

71. Morton J. Elrod to Chancellor Edward C. Elliott, Madison, Wisconsin, 14 January 1916, MJE, S2, B2, F1.

72. Elrod to Elliott, 7 June 1916, MJE, S5, B29, F9.

73. Morton J. Elrod to Judge W. M. Bickford, 3 July 1916, MJE, S5, B29, F9.

74. M. D. Baldwin to Morton J. Elrod, 26 July 1916, MJE, S5, B29, F9.

75. Morton J. Elrod to M. D. Baldwin, 27 July 1916, MJE, S5, B29, F9.

76. Morton J. Elrod to Dr. Josiah J. Moore, 28 August 1916, MJE, S2, B2, F2. No response in the Elrod papers.

77. Morton J. Elrod, typescript report concerning the Lake and the sites for fish work, 1918; W. M. Bickford to Morton J. Elrod, 1 June 1916, authorizing the study; and Morton J. Elrod to W M. Bickford, 22 May 1916, outlining plans for the study, all MJE, S5, B29, F10; see also Morton J. Elrod to W. M. Bickford, 26 June 1917, MJE, S2, B2, F3.

78. Morton J. Elrod to H. B. Ward, 12 February 1916; Morton J. Elrod to Chancellor Elliott, 14 April 1916; and Chancellor Edward C. Elliott to Morton J. Elrod, 24 April 1916, setting the budget at $12,000 but obviously meaning $1,200, all MJE, S5, B29, F10; see also Morton J. Elrod, "Biological Station, Outline for 1916," 1916, MJE, S5, B23, F6.

79. Morton J. Elrod to Edward C. Elliott, "Report . . . of the Biological Station," 1916, with a request for 1917, MJE, S5, B29, F10.

80. Morton J. Elrod, "Flathead Lake Report," 1916, MJE, S5, B30, F1. See also Morton J. Elrod to President E. O. Sisson and Chancellor Elliott, "Preliminary Report of the Work at the Biological Station on Yellow Bay, Flathead Lake, Season 1918," 12 August 1919, RG1, PO, S15, B30, F "Biological Station, 1900–1929."

81. Elrod, "Flathead Lake Report," 1916, MJE, S5, B30, F6.

82. Morton J. Elrod to Chancellor Edward C. Elliot, 2 March 1917, MJE, S5, B30, F2.

83. "Protozoa," at http://en.wikipedia.org/wiki/Protozoa.

84. J. M. Aldrich, Associate Curator, Smithsonian Institution, to Morton J. Elrod, 20 August 1924, MJE, S4, B14, F8. He had not found a publisher by 1924.

85. Morton J. Elrod to Acting President Scheuch, 13 March 1917, MJE, S5, B30, F3.

86. Morton J. Elrod to F. C. Scheuch, 11 April 1917; Edward C. Elliott to Morton J. Elrod, 6 May 1917; and Morton J. Elrod to F. C. Scheuch, 1 June 1918, all MJE, S5, B30, F3.

87. Morton J. Elrod to W. M. Bickford, 8 June 1917, MJE , S5, B30, F3.

88. Elrod to Bickford, 26 June 1917, MJE, S2, B2, F3.

89. Morton J. Elrod, handwritten note on letter from A. S. Pearse to members of the Research Committee on Fish and Fisheries, Ecological Society of America, 4 April 1918, MJE, S3, B8, F1.

90. Morton J. Elrod to Dave Elrod, 9 August 1917, MJE, S2, B2, F3.

91. Nelson Story, Jr., to W. M. Bickford, 12 June 1917; J. L. DeHart to W. M. Bickford, 15 June 1917; M. D. Baldwin to W. M. Bickford, 25 June 1917; W. M. Bickford to Morton J. Elrod, 9 July 1917; and Morton J. Elrod to W. M. Bickford, 9 August 1917, all MJE, S2, B2, F3.

92. Morton J. Elrod to Walter M. Bickford, 11 August 1917, emphasis supplied, MJE, S2, B2, F3.

93. Elrod, "Flathead Lake Report," 1916, MJE, S5, B30, F6.

94. Morton J. Elrod to Judge Bickford, 16 August 1917, MJE, S2, B2, F3.

95. H. M. Parchen to Morton J. Elrod, 23 May 1917; and Morton J. Elrod to H. M. Parchen, 11 August 1917, both MJE, S2, B2, F3.

96. Morton J. Elrod to President Edward O. Sisson, 22 January 1918, MJE, S5, B30, F4.

97. For different listings, see Morton J. Elrod, handwritten list (no date, but 1918), MJE, S2, B1, F10 and MJE, S5; B30, F4; and Morton J. Elrod, typescript "List of papers and articles submitted for publication" (no date but 1921), MJE, S5, B30, F12; also "Numbered Publications," clipped file, March 1934, RG1, PO, S19, B46, F "Publications, Procedures for Distribution to State Board"; and the composite listing in the appendix herein.

98. Morton J. Elrod to President E. O. Sisson, 8 and 29 March 1918; and Edward C. Elliott to E. O Sisson, 5 April 1918, all MJE, S5, B30, F4.

99. Elrod referred to a chancellor's letter, so he obviously received and read it, as discussed in his letter of 2 March 1917 to Elliott (MJE, S5, B30, F2), but not found in the Elrod papers or the president's office records.

100. Elliott to Sisson, 5 April 1918, MJE, S5, B30, F4; and see Beard, "Morton J. Elrod," Section 3, 4–5.

101. E. O. Sisson to Professor Elrod, 19 January 1918, RG1, PO, S15, B30, F "Biological Station, 1900–1929."

102. P. W. Graff to Pres. Sisson, 2 April 1918, RG1, PO, S15, B30, F "Biological Station, 1900–1929."

103. M. J. Elrod, "The Biological Station for 1918" (no date, but February–April 1918), RG1, PO, S15, B30, F "Biological Station, 1900–1929." See also Beard, "Morton J. Elrod," Section 3, 4–5.

104. Elrod, "Preliminary Report of . . . Work at the Biological Station . . . , Season 1918," 12 August 1919, RG1, PO, S15, B30, F "Biological Station, 1900–1929."

105. "Diatom," http://en.wikipedia.org/wiki/Diatom.

106. Burlington Moore, Editor-in-Chief, *Ecology,* to Morton J. Elrod, 19 August 1920, commending "Evaporation at Flathead Lake" for the value of the data and suggesting revision of the article, MJE, S5, B32, F9. No evidence of a revision.

107. T. C. Frye, Director, to President E. O. Sisson, 28 February 1919, RG1, PO, S15, B30, F "Biological Station, 1900–1929." See also Beard, "Morton J. Elrod," Section 3, 4–5.

108. Eric Gunderran to Morton J. Elrod, 5 June 1919, MJE, S5, B30, F9.

109. Thas Spaulding, Acting Dean, to Morton J. Elrod, 7 July 1919, MJE, S5, B30, F9.

110. "Great Fire of 1910," at http://en.wikipedia.org/wiki/Great_Fire_of_1910. See also Gil Gale, "Fire Ferocity: Wildfires of the Past," *Montana Naturalist* (Winter 2014–2015), 9; and Dan Miller and Stan Cohen, *Big Burn: The Northwest's Great Forest Fire of 1910* (Missoula: Pictorial Histories Pub. Co., 1978), passim.

111. Edward C. Elliott to Morton J. Elrod, 26 July 1919, MJE, S5, B30, F9.

112. Morton J. Elrod, "Report of the . . . Biological Station," 1921, MJE, S5, B30, F11.

113. "Shy of Funds; Station Closes," *Missoulian* (20 June 1920), MJE, S5, B31, F21.

114. Clapp, "Narrative," ch. 2, 10–11, and ch. 6, 46–50; and Merriam, *History,* 55, 63–64.

115. Elrod, "Report of the . . . Biological Station," 1921, MJE, S5, B30, F11.

116. Morton J. Elrod and Francis Ross, "The Fishes of Flathead Lake," 1921, MJE, S5, B30, F11.

117. Morton J. Elrod, handwritten fragment, 1921, MJE, S5, B30, F11.

118. Robert Oslund, Woods Hole, to Morton J. Elrod, 24 July 1921, MJE, S2, B2, F6.

119. Clapp, "Narrative," ch. 6, 41. Rumor circulated that either the board or faculty forced Sisson to leave, but he wrote years later to deny that the State University had become a "graveyard for presidents," *Missoulian* (15–27 January 1940), various excerpts, RG1, PO, S4, B167, F "State Board Hearing, 1940."

120. Morton J. Elrod to Emma, 26 April 1921, MJE, S2, B2, F6.

121. Morton J. Elrod to Emma and Mary, 28 March 1922, MJE, S2, B2, F7. See also Morton J. Elrod, fragment (no date, but 1922), MJE, S4, B19, F7, indicating that Superintendent J. R. Eakin proposed the "educational work" among park visitors to President Clapp, Professor J. R. Kirkwood, Elrod, and probably R. J. Young, based on a successful experiment in Yosemite National Park.

122. Morton J. Elrod to "Folks" [reference to Emma, Mary, and her husband, William Ferguson], 21 August 1922, MJE, S2, B2, F7.

123. Morton J. Elrod to Emma, 2 August 1922, MJE, S2, B2, F7.

124. Clapp, "Narrative," ch. 7, 12; and Morton J. Elrod to J. R. Eakin, 18 May 1923, MJE, S4, B14, F17.

125. Elrod, "Introduction" (1926–30), MJE, S4, B15, F12. See also Morton J. Elrod, "A Week at Brown's Pass," 1911, MJE, S4, B15, F6; and Morton J. Elrod, "Brown's Pass—Bowman Lake Trail," 1911, MJE, S4, B18, F8.

126. Elrod, "Introduction" (1926-30), 17, on the cameras, MJE, S4, B15, F12.

127. Morton J. Elrod, *Some Lakes of Glacier Park* (Washington, DC: Department of the Interior, 1912), 29 (available on line at http://archive.org /details/somelakesofglaci01elro), with 14 photographs and descriptions of 14 lakes. For a description of his methods, see Morton J. Elrod, "Report of the Park Naturalist, Glacier National Park, 1926," GBG, R42; and Morton J. Elrod, "Photography in the Park" (no date but in folder marked 1912), MJE, S4, B18, F12, mentioning incidents of the 1906 and 1910 excursions.

128. Morton J. Elrod to Emma, 5 September 1925, MJE, S2, B2, F10. Also Morton J. Elrod, "Dredging on Lake Louise" (no date but 1910–14), MJE, S4, B19, F5; and Morton J. Elrod, "The Lakes of Glacier National Park in Relation to Fish" (no date but1910–15), MJE, S4, B19, F8. See as well Morton J. Elrod, "Iceberg Lake" (no date, but 1910–15), *Mountaineer,* 43–51; and Morton J. Elrod, handwritten fragment describing Iceberg Lake, probably 1910–14, both MJE, S4, B17, F3. See also Elrod, "Report of the Park Naturalist," 1926, GBG, R42, on dredging McDermott and Josephine Lakes; and Elrod, "Montana's New National Park," *Missoulian* (21 July 1910), R39, on fish in the lakes.

129. Robert G. Athearn, "The Tin Can Tourists' West," in *Montana and the West: Essays in Honor of K. Ross Toole,* edited by Rex C. Myers and Harry W. Fritz (Boulder, CO: Pruett Publishing Company, 1984), 109.

130. Harper, "Conceiving Nature," 14, 92.

131. Athearn, "Tin Can Tourists' West," 108–109; Harper, "Conceiving Nature," 18; and Elrod, "Montana's New National Park," *Missoulian* (21 July 1910), R39.

132. Athearn, "Tin Can Tourists' West," 109. For infrastructure development, see Diettert, *Grinnell's Glacier,* ch. 9; H. A. Noble, Manager, Glacier Park Hotel Company, to Mr. Geo. Bird Grinnell, 12 April 1919, GBG, R42, urging Grinnell to consider a booklet and an article on the park; and W. R. Logan, Superintendent, to Geo. Bird Grinnell, 23 May 1911, GBG, R42.

133. H. A. Noble, General Passenger Agent, Great Northern Railway, to Morton J. Elrod, 4 June 1912, MJE, S4, B14, F2. Noble became Manager of the Glacier Park Hotel Company in later years.

134. For the quotation and details, see Buchholz, *Man in Glacier,* ch. 5, at http://www.nps.gov/history/history/online_books/glac2/; Logan to Grinnell, 23 May 1911, GBG, R42; "Senator J. M. Dixon Gets Back Home" and "Glacier Park Will Soon Be Opened," *Missoulian* (21 July 1910), both R39. For the Great Northern Railway's contributions of publicity and infrastructure, see Peterson, *Call of the Mountains,* ch. 2.

135. The author recalls camping trips in the late 1930s to the Avalanche Creek campground.

136. Buchholz, *Man in Glacier,* ch. 5, at http://www.nps.gov/history/history /online_books/glac2/; National Park Service, *Glacier: Through the Years in Glacier National Park,* ch. 3, at http://www.nps.gov/history/history/online_books /glac/chap3.htm; and C. W. Guthrie, *Great Northern Railway and Glacier National Park* (Helena: Far West, 2004), 43.

137. Morton J. Elrod to Emma and Mary, 28 March 1922, MJE, S2, B2, F7; and see Noble to Elrod, 4 June 1912, MJE, S4, B14, F2.

138. Morton J. Elrod, to Folks (referring to Emma, daughter Mary, and son-in-law William), 9 July 1922, MJE, S2, B2, F7.

139. Morton J. Elrod to Emma and Mary, 20 July 1922, MJE, S2, B2, F7. See also National Park Service, "Glacier," ch. 3, at http://www.nps.gov/history/history/online_books/glac/chap3.htm. For discussion of early tourism in the west, see Athearn, "Tin Can Tourists' West," passim.

140. Elrod, *Elrod's Guide,* 116; see also Buchholz, *Man in Glacier,* ch. 5, at http://www.nps.gov/history/history/online_books/glac2/. See also Peterson, *Call of the Mountains,* although Peterson did not include Dennett, he included Charlie Russell and most other prominent artists and writers who visited the park in the early days.

141. Morton J. Elrod to Emma, 16 July 1924, MJE, S2, B2, F9.

142. J. R. Eakin, Superintendent, to Morton J. Elrod, 1 August 1922, MJE, S2, B2, F7. See also J. R. Eakin to Morton J. Elrod, 6 December 1922, trail descriptions and pieces on movement of Grinnell Glacier, MJE, S4, B14, F6.

143. Morton J. Elrod to Emma, Mary and William, 28 July 1922, MJE, S2, B2, F7.

144. Eakin to Elrod, 1 August 1922, MJE, S2, B2, F7.

145. Morton J. Elrod to Emma, 2 August 1922, MJE, S2, B2, F7.

146. Morton J. Elrod to Folks (referring to Emma, daughter Mary, and son-in-law William), (undated but early August 1922), MJE, S2, B3, F12.

147. For another account of the Grizzly encounter, see Morton J. Elrod, typescript, "Mammals of Glacier Park" (no date but 1925 or 1926), identifying Ranger Lou Sarratt, a temporary appointee who taught at Northwestern University, MJE, S4, B15, F1.

148. Morton J. Elrod to Emma, 12 August (no year but 1922); Morton J. Elrod to Emma, 26 July (no year but 1924), both MJE, S2, B3, F12. And see E. J. Roessner, Managing Editor, NEA Service, to Morton J. Elrod, 25 September 1925, a "new service" failed so he returned manuscripts, MJE, S2, B2, F10. See as well Jas. P. Brooks, Asst. Supt., to Morton J. Elrod, 26 July 1924, listing several typed articles; and Morton J. Elrod, excerpts from newspapers (various dates), all MJE, S4, B14, F8.

149. Morton J. Elrod to Emma, "Friday" (no date but 29 June 1928), MJE, S2, B3, F1.

150. Morton J. Elrod to Emma, 30 August 1928, MJE, S2, B3, F1.

151. Elrod to Emma, 2 August 1922, MJE, S2, B2, F7.

152. Morton J. Elrod to Folks (referring to Emma, daughter Mary, and son-in-law William), 21 August 1922, MJE, S2, B2, F7.

153. Elrod, *Elrod's Guide,* 258, originally published in 1924, revised edition in 1930. The standard edition sold for $1.00 and the deluxe for $2.00 or $2.50.

154. H. A. Noble to Morton J. Elrod, 12 November 1922, MJE, S4, B14, F6.

155. Discussion based on the 1930 edition. See also Robert F. Haynes, "About Dr. Morton Elrod, Park's First Naturalist," *Hungry Horse News* (14 December 1956), MJE, S4, B14, F21; and Morton J. Elrod, "Photography in the Park," typescript fragment (undated but mid-1920s), taking photographs in the Park, MJE, IV, B18, F12. For examples of updated material, see C. P. Fordyce, Camping Editor, *Outdoor Life,* to Morton J. Elrod, 31 October 1929, MJE, S4, B14, F14; and C. J. Kraebel to Morton J. Elrod, 19 December 1924, MJE, S4, B14, F8. See also James Ridler to Morton Elrod, 6 February 1928; Morton J. Elrod to Emma, 12 July 1928; and Morton J. Elrod to Emma, 8 July 1928, all MJE, S2, B3, F1.

156. Elrod, *Elrod's Guide,* 28, 98, 116, 119, 130, 137, 250–55. On Schultz, see "James Willard Schultz" at http://en.wikipedia.org/wiki/James_Willard_Schultz; and on Grover, see "Oliver Dennett Grover," at http://en.wikipedia.org/wiki/Oliver_Dennett_Grover.

157. Elrod, *Guide,* 232–34.

158. Morton J. Elrod to Mr. McGillis, GN Railway, St. Paul, 24 March 1930, MJE, S4, B14, F15; and J. R. Eakin to Morton J. Elrod, 18 July 1924, MJE, S4, B14, F8.

159. Morton J. Elrod to Emma, 15 July 1924, MJE, S2, B2, F9; Morton J. Elrod to Emma, 13 August 1925, MJE, S2, B2, F10; and Morton J. Elrod to Emma, 10 July 1928, MJE, S2, B3, F1.

160. Elrod to Emma, 15 July 1924, MJE, S2, B2, F9; and Morton J. Elrod, "Sales Report, First Edition of Elrod's Guide," 25 July 1930, MJE, S4, B15, F14.

161. See, for example, G. W. Noffsinger, Park Saddle Horse Company, to Morton J. Elrod, 27 March 1930, a company trail guide using Elrod's descriptions, MJE, S4, B14, F15. See also Morton J. Elrod, typescript, "Some Lakes of Glacier National Park" (no date but about 1911), MJE, S4, B17, F4–5; Elrod,

Some Lakes of Glacier Park, 1912, the published pamphlet (available on line at http://archive.org/details/somelakesofglaci01elro). And see Morton J. Elrod, handwritten and typescript, "Some Additional Lakes in Glacier National Park," 1919, MJE, S4, B17, F7; Morton J. Elrod, "Five Day North Circle Trip," 1926, MJE, S4, B15, F19; Morton J. Elrod, "The Passes of Glacier National Park" (no date but early 1920s), MJE, S4, B18, F9–10; Morton J. Elrod, "Walks About Granite Park" (no date but probably early 1920s), MJE, S4, B16, F2; Morton J. Elrod, "Untraveled Trails of Glacier National Park," 1922, MJE, S4, B20, F5.

162. Morton J. Elrod to Emma, 11 June 1925, MJE, S2, B2, F10.

163. Elrod to Emma, 15 July 1924, MJE, S2, B2, F9; and see Elrod to Emma, 11 June 1925, MJE, S2, B2, F10.

164. Morton J. Elrod to Emma, 4 August 1925, MJE, S2, B2, F10, the earlier visit in 1910.

165. Morton J. Elrod to Emma, 10 and 29 July, and 2 and 30 August 1928, all MJE, S2, B3, F1.

166. Elrod to Emma, 29 July 1928, MJE, S2, B3, F1.

167. Elrod to Emma, 2 August 1929, MJE, S2, B3, F1.

168. James Ridler, Superintendent of Stores, Great Northern Company, to Morton J. Elrod, 10 February 1930, MJE, S4, B14, F15.

169. See handwritten or typed chapters in the Elrod papers, such as Morton J. Elrod, "Effects of Wind" (no date but mid- to late 1920s); Morton J. Elrod, "Plant Injuries from Snow" (no date but the mid- to late 1920s); and Morton J. Elrod, "Avalanches" (no date but mid- to late 1920s), all MJE, S4, B15, F13.

170. Morton J. Elrod to Mr. McGillis, GN Railway, St. Paul, 24 March 1930, MJE, S4, B14, F15.

171. O. J. McGillis, General Advertising Agent, G.N., to Morton J. Elrod, 14 July and 8 August 1930, both MJE, S4, B14, F15.

172. Morton J. Elrod to Emma, 14 July 1927, MJE, S2, B2, F12.

173. Morton J. Elrod to Emma, "Friday" (probably 29 June 1928), MJE, S2, B3, F1.

174. Morton J. Elrod, 24 handwritten pages (no date but late 1920s), listing the packages sent to people in various states, orders from $0.50 to $10.00, totaling roughly $140.00, MJE, S7, B37, 17.

175. Several citations of probable chapters or sections appear above and below. See also Elrod to Badger, 5 July 1917, "the manuscript for a proposed book about Glacier National Park," not found in the Elrod papers, MJE, S2, B2, F9.

176. Elrod, typescript fragment about old-timers he had met (late 1920s), and citations above.

177. Morton J. Elrod to Emma, 12 July 1926, MJE, S2, B2, F11; see also Elrod, "Report of the Park Naturalist, 1926," GBG, R42; and Peterson, *Call of the Mountains,* an account of Elrod and Grinnell in the Many Glacier Hotel.

178. George Bird Grinnell to Morton J. Elrod, 1 September 1926, on the hunting trip, MJE, S2, B2, F11; also M. W. Beacom to George B. Grinnell, 30

April 1917, and Grinnell to Yard, 27 October 1927, both GBG, R42. See also Diettert, *Grinnell's Glacier*, ch. 4.

179. Grinnell to Elrod, 1 September 1926, MJE, S2, B2, F11. And see Morton J. Elrod, typescript, "Grinnell Glacier" (no date but 1928–29), photographs depicting the changes to the glacier, MJE, S4, B16, F10–12.

180. Grinnell to Elrod, 1 September 1926, MJE, S2, B2, F11.

181. Morton J. Elrod, "To Determine the Rate of Movement of Grinnell Glacier" (no date but 1922), including notes on various visits, MJE, S4, B16, F10–12; Morton J. Elrod, "Elrod Describes Grinnell Glacier," *Missoulian* (13 May 1923), 1, 8, MJE, S4, B17, F2; W. N. Noffsinger, Park Saddle Horse Company, to Morton J. Elrod, 16 October 1922, on the stakes, MJE, S4, B14, F6; and Elrod, "Report of the Park Naturalist," 1926, GBG, R42.

182. Elrod, "The Glaciers of Glacier National Park" (1926–29), MJE, S4, B16, F1. See also Elrod, "Report of the Park Naturalist," 1926, GBG, R42; and Morton J. Elrod, "Trip to Sperry Glacier," newsprint, no source (1928–31), MJE, S4, B20, F1.

183. Morton J. Elrod, handwritten and typescript, "Museum (Clements) Glacier" (no date but 1931), MJE, S4, B17, F14; and Morton J. Elrod, "Two Oceans Glacier," in Elrod, "The Glaciers of Glacier National Park" (1926–29), MJE, S4, B16, F1.

184. C. J. Kraebel to Morton J. Elrod, 4 September 1925, MJE, S4, B14, F9.

185. Morton J. Elrod, "Blackfoot Ice Mass, Glacier Park, Receding," *Missoulian* (21 March 1926), 1, 12, MJE, S4, B14, F21.

186. For the quotations, see Morton J. Elrod, "Harrison Glacier," in Elrod, "The Glaciers of Glacier National Park" (1926–29), MJE, S4, B16, F1. See also Morton J. Elrod to Emma, 5 September 1925, MJE, S2, B2, F10; and Elrod, "Blackfoot Glacier," 1911, MJE, S4, B15, F4.

187. Elrod, "Blackfoot Ice Mass, Glacier Park, Receding," 1926, MJE, S4, B14, F21.

188. Elrod, "Report of the Park Naturalist," 1926, GBG, R42; and J. W. Emmert to Morton J. Elrod, 15 April 1952, MJE, S2, B3, F11.

189. Morton J. Elrod, "The Matterhorn," handwritten, 24 May 1909, Cosmos Club paper, MJE, S8, B37, F8; and Morton J. Elrod, handwritten, "Modern Mountaineering" (no date but 1910s), Cosmos Club paper, MJE, S8, B37, F10.

190. Morton J. Elrod, "An Electric Storm in Montana," handwritten and typescript, *Youth's Companion* (2 November 1911); also *Kaimin* (1911), both MJE, S3, B7, F4.

191. Elrod, handwritten and typescript, "Attempt on St. Nicholas" (no date but 1906), MJE, S4, B19, F9 and 10, respectively.

192. On Clyde, see "Norman Clyde" at http://en.wikipedia.org/wiki /Norman_Clyde. See also Morton J. Elrod, handwritten, "Norman Clyde," 1928, MJE, S4, B15, F8; and Norman Clyde to Morton J. Elrod, 18 August 1924, MJE,

S4, B14, F8. See as well Morton J. Elrod, "Conquering Glacier Park Peaks," *Missoulian* (3 February 1929), Leopold Seetholder, a Swiss guide, and Clyde, MJE, S4, B15, F11.

193. Conrad Wellen to Morton J. Elrod, 2 July 1930, MJE, S4, B14, F14. Also "Mount Saint Nicholas," on Reverend Conrad Wellen, Havre, Montana, at http://en.wikipedia.org/wiki/Mount_Saint_Nicholas. And see Morton J. Elrod, "Paul J. Moody," handwritten (no date but after 1928), Wellen's ascent of St Nicholas, MJE, S4, B5, F18.

194. W. G. Peters, Associate Highway Engineer, Transmountain Highway, to Morton J. Elrod, 7 August 1928, MJ, S4, B14, F13.

195. Morton J. Elrod, "Mountain Climbing in Glacier," *Missoulian* (27 January 1929), handwritten and typescript, MJE, S4, B14, F22 and B17, F11, respectively.

196. G. M. Kilbourn to Morton J. Elrod, 28 August 1928, MJE, S4, B14, F1.

197. Morton J. Elrod, handwritten fragment (no date), trip to Kintla, Kootenai, Louise, and other lakes for dredging, probably in 1910 or 1914, MJE, S4, B15, F12.

198. Morton J. Elrod, "Guides of Glacier Park Are Real Men, Have Arduous Tasks," *Missoulian* (24 January 1926), 1, 8, MJE, S4, B14, F23.

199. The incidents relayed in Morton J. Elrod to Emma, Mary, and William, 23 and 28 July 1922, MJE, S2, B2, F7; and to Emma, 8, 12, 14, and 23 July 1925, MJE, S2, B2, F10.

200. Morton J. Elrod, "Conservation" (no date, but after 1924, probably 1925), MJE, S4, B15, F9.

201. Morton J. Elrod, "Department of Biology" (undated, but 1916), MJE, S5, B23, F6.

202. National Park Service, "Glacier National Park Census of The Wild Game," numbers in parentheses refer to estimated totals (no date but 1924), MJE, S4, B15, F2.

203. Morton J. Elrod, "Mammals of Glacier Park," typescript (no date, but 1925 or 1926), MJE, S4, B15, F1. See also Morton J. Elrod to Emma, 22 August 1925, referring to article on the "Bison herd," *Journal of Mammalogy,* and a request for one on "game animals of Glacier Nat. Park," MJE, S2, B2, F10.

204. "Glacier National Park Census" (1924), MJE, S4, B15, F2.

205. Morton J. Elrod, "Coyote," handwritten and typescript (no date but 1925 or 1926), MJE, S4, B15, F1.

206. Morton J. Elrod to Emma, 16 and 28 July 1924, MJE, S2, B2, F9; Elrod, "The Inter-Mountain Educator," *Montana Education,* 15–16; also full page advertisements by all four state institutions in the *Educator* (ten per year); Morton J. Elrod to Melvin A. Brannon, 11 March MJE, S3, B6, F9; and Elrod, "Play Fair, Boys," *Inter-Mountain Educator,* 216.

207. John Hartshorn, Vice President, Black Stewart Coal Company, to Morton J. Elrod, 18 December 1924, MJE, S2, B2, F9.

208. John Hartshorn to Mrs. M. J. Elrod, telegram, 10 December 1924, MJE, S2, B2, F9.

209. John Hartshorn to Emma, 19 December 1924, MJE, S2, B2, F9.

210. John Hartshorn to Morton J. Elrod, 7 January 1925, MJE, S2, B2, F10.

211. Morton J. Elrod to Emma, 11 June 1925, MJE, S2, B2, F10. See also Morton J. Elrod to Emma, 16 July 1927 (or 1928), MJE, S2, B3, F1.

212. Morton J. Elrod to Emma, 9 July 1925, MJE, S2, B2, F10.

213. John Hartshorn to Mort and Emma, 15 June 1925, MJE, S2, B2, F10.

214. Elrod to Emma, 11 June, to Emma, 8 July, 8 July (second letter), 14 July, and 16 August 1925, all MJE, S2, B2, F10.

215. J. R. Eakin to Morton J. Elrod, 27 February 1923; and H. A. Noble to Morton J. Elrod, 24 February 1923, both MJE, S5, B14, F7.

216. Morton J. Elrod to J. R. Eakin, 18 May 1923, MJE, S4, B14, F7.

217. M. J. Elrod to President Clapp, 7 August 1923, RG1, PO, S15, B30, F "Bacteriology."

218. Harriett H. Roter, Clerk, Educational Headquarters, National Park Service, to Morton J. Elrod, 26 July 1926, MJE, S4, B14, F10. See also Elrod, *Elrod's Guide*, 232–34.

219. Morton J. Elrod, "Report," 1925, MJE, S4, B19, F1; "Report," 1926, MJE, S4, B19, F2; "Report," 1927, MJE, S4, B19, F3; and "Suggestions to Increase the Service," 1927, MJE, S4, B19, F4. Also Morton J. Elrod, "Glacier National Park, Ranger Naturalist Service" (no date but mid-1920s), MJE, S4, B19, F4; for numerous notes on signage, fishing, document sales, etc., MJE, S4, B19, F4; also Morton J. Elrod, "Preparing for Increased Travel," "The Season's Travel," and "Distribution of Travel" (no dates but 1922–24); and "Report on Lecturer Attendance . . . 1928" and "Field Trips," all MJE S4, B19, F7; and Reports of Ranger Naturalists to M. J. Elrod (no dates but mid- to late 1920s), MJE, S4, B19, F8, including R. J. Young, who became assistant director of the Biological Station.

220. Elrod to Emma, 12 July 1926, MJE, S2, B2, F11.

221. Morton J. Elrod to Emma, "Friday" (no date but probably 29 June 1928); and Morton J. Elrod to Emma, 10 July and 30 August 1928, all MJE, S2, B3, F1.

222. W. H. Mills to Morton J. Elrod, 3 March 1926; and Morton J. Elrod to W. H. Mills, 16 March 1926, both MJE, S4, B14, F10.

223. Elrod, "The Relationship of the People," 1923, MJE, S4, B18, F17; Morton J. Elrod, "Educational Building in Glacier National Park Would Give Visitors Better Understanding of Wonderland, Says Elrod," *Great Falls Tribune* (2 September 1923), 3, 9; and several other press excerpts, MJE, S4, B14, F21.

224. C. J. Kraebel to Morton J. Elrod, 19 December 1924, MJE, S4, B14, F8.

225. Charles J. Kraebel to Morton J. Elrod, 26 April 1926, MJE, S4, B14, F10.

226. G. S. Scott, Chief, Division of Appointments, to Morton J. Elrod, 31 May 1928; and J. R. Eakin to Morton J. Elrod, 15 July 1928, both MJE, S4, B14, F13.

227. The discussion draws on Morton J. Elrod to Emma, 29 June 1928, 2, 4, 8, 10, 12, 16, 21, and 25 July, and 30 August 1928, all MJE, S2, B3, F1; C. J. Kraebel to Morton J. Elrod, 7 December 1925, MJE, S4, B14, F9; and R. R. Vincent, Asst. Supt., to Morton J. Elrod, 20 August 1928, MJE, S4, B14, F13.

228. G. E. Scott, Chief, Division of Appointments, to Morton J. Elrod, 14 April 1929; and J. R. Eakin to Morton J. Elrod, both, MJE, S4, B14, F14.

229. George Bird Grinnell to Morton J. Elrod, 3 June 1926, MJE, S4, B14, F10; H. A. Noble to Morton J. Elrod, 29 May 1926, MJE, S4, B14, F11; and Morton J. Elrod, various materials, typewritten and printed, "Glacier National Park Museum Society, Organized in January 1926," 1926, MJE, S4, B18, F6 and MJE, S4, B17, F13.

230. Morton J. Elrod, "General Plan and Suggestions for Use of Museum and Student Building" (no date but 1926 or 1927), MJE, S4, B17, F12.

231. McGillis to Elrod, 14 July 1930 and 8 August 1930, both MJE, S4, B14, F15.

232. Noffsinger to Elrod, 27 March 1930, MJE, S4, B14, F15.

233. George C. Ruhle, Park Naturalist, to Morton J. Elrod, 17 February 1930, MJE, S4, B14, F15.

234. Ralph C. Teall, Ranger Naturalist, to G. C. Ruhle, Park Naturalist, 23 July 1930, MJE, S4, B14, F15.

235. J. R. Eakin to Morton J. Elrod, 22 May 1930, MJE, S4, B14, F15.

236. J. R. Eakin to Morton J. Elrod, 5 July 1930, MJE, S4, B14, F15.

237. Melvin A. Brennan to Morton J. Elrod, 27 March 1928, MJE, S2, B2, F5; Morton J. Elrod, "Three Memoranda Concerning the Fish Study on Flathead Lake in 1928 and 1929" (undated but 1928), MJE, S5, B30, F14.

238. Elrod, "Introduction," 1926–30, MJE, S4, B15, F12; see also his notes in MJE, S4, B18, F7.

Chapter 5

1. Asst. Business Manager, State University, to A. J. Casey, Western Supply Co., Kalispell, 1 June 1926, quoting Elrod, RG1, PO, S15, B30, F "Biological Station, 1900–1929."

2. Morton J. Elrod to President Clapp, 30 March 1928, RG1, PO, S15, B30, F "Biological Station, 1900–1929."

3. C. H. Clapp, President, to Chancellor M. A. Brannon, 26 January and 20 February 1928; and Chancellor Melvin A. Brannon to President C. H. Clapp, 25 January and 24 February 1928, RG1, PO, S15, B30, F "Biological Station, 1900–1929"; and Melvin A. Brannon to Morton J. Elrod, 27 March 1928, MJE, S2, B2, F5.

4. Melvin Amos Brennan, at https://www.beloit.edu/archives/history/presidents/melvin_brannon/.

5. Clapp, "Narrative," ch. 7, 66–71; The Phi Beta Kappa Association, "Petition," 1929, RG1, PO, S15, B26, F "ARCHIVES: Informational"; typescript of a State Supreme Court decision, *ex rel Frances D. Jones vs. The State Board of Examiners,* decided on 20 February 1926, initiated by Brannon to challenge successfully the board of examiners, RG1, PO, S3, B16, F "1919, 1923, 1924, Legislative Material"; also "Senate Bill No.1, 1933, RG1, PO, S3, B16, F "Legislative, 1933, Millage data"; Clapp, "Remarks on Centralized Systems," RG1, PO, S5, B6, F "National Association of State Universities, 1910–1951"; Merriam, *History,* ix–x, 66, 69; Lucille J. Armsby, to President McFarland, 5 April 1951, on Brannon's resignation, RG1, PO, S19, B52, F "Duplication Study, 1951–51"; and James D. Graham's minority opinion in "Report of the Montana Commission on Higher Education," 26 September 1944, RG1, PO, S19, B52, F "Duplication—Melby Attempts to Change Unit Functions, 1944–5."

6. Chancellor Melvin A. Brannon to President C. H. Clapp, 7 May 1928, RG1, PO, S15, B30, F "Biological Station, 1900–1929."

7. Elrod, "Three Memoranda" (undated but 1928), MJE, S5, B30, F14.

8. Morton J. Elrod, Report on "the general problem of work," 13 May 1928, RG1, PO, S15, B30, F "Biological Station, 1900–1929."

9. Elrod, "Three Memoranda" (undated but 1928); and Fish and Game Commission reports, *Montana Wildlife* 2, no. 1 (June 1929), 3–15, both MJE, S5, B31, F24. See also Morton J. Elrod, "Plan for Investigation on Fishes of Flathead Lake and Georgetown Lake for the Station for 1929," 1929, MJE, S5, B30, F14.

10. Morton J. Elrod, "Eulogy" and "Statement," 1928, both MJE, S5, B28, F1; and "Kirkwood, Joseph Edward (1872–1928)," *JSTOR PLANT SCIENCE,* at http://plants.jstor.org/person/bm000052569.

11. Elrod, "Three Memoranda," (undated but 1928), MJE, S5, B30, F14.

12. Morton J. Elrod to Emma, 29 July 1928, MJE, S2, B3, F1.

13. President C. H. Clapp to Chancellor M. A. Brannon, 12 January and 18 July 1929, the latter including the1928 report; and Chancellor M. A. Brannon to President C. H. Clapp, 16 January 1928, both RG1, PO, S15, B30, F "Biological Station, 1900–1929"; and Reports on Work in 1928, *Montana Wildlife* (June 1929), 3–15, MJE, S5, B31, F24.

14. Robert H. Hill, State Game Warden, to Hon. Melvin A. Brannon, Chancellor, 22 April 1929; Melvin A. Brannon to President C. H. Clapp, 19 June 1929; President C. H. Clapp to Chancellor M. A. Brannon, 23 June 1929; and Morton J. Elrod, "Recommendations made to Mr. T. N. Marlowe, chairman, Montana Fish and Game Commission, . . . summer 1929"; all RG1, PO, S15, B30, F "Biological Station, 1900–1929"; Morton J. Elrod, "Concerning Dr. Young's Suggestions" (no date but 1929); Morton J. Elrod to T. N. Marlowe, Chairman, Montana Fish and Game Commission (no date but June 1929); and Morton J. Elrod, "Plan for Investigation on Fishes of Flathead lake and Georgetown Lake for the Station for 1929," 1929, all MJE, S5, B30, F14.

15. E. L. Wickliff to Thomas N. Marlowe, 7 April 1930, MJE, S3, B6, F10.

16. T. N. Marlowe to Morton J. Elrod, 14 April 1930, MJE, S2, B6, F10.

17. Chancellor Melvin A. Brannon to Morton J. Elrod, 8 May 1930, MJE, S3, B6, F10.

18. Except as specifically noted, the following analysis draws on the reports on work in 1928 at Flathead Lake, *Montana Wildlife* (June 1929), 3–15, MJE, S5, B31, F24; Morton J. Elrod, "Vertebrates of Flathead Lake" (no date but 1930 or 1931), 1928–29, reports of the Lake Superior whitefish and some sockeye salmon, amphibians, and reptiles, MJE, S5, B33, F9; and "Flathead Lake" (no date but 1929–30), MJE, S5, B32, F2. Also see Morton J. Elrod, "Vertebrates of Flathead Lake" (same document as listed earlier in this note but specifically dated February 1930); and Elrod's reports on earlier work, especially Morton J. Elrod, "The Fishes and the Fish Food of Flathead Lake," report of work in 1915–16 and 1918 with Robert Oslund and Maurice Pace, Superintendent of Polson; and "Rise and Fall and Drainage and Evaporation of Flathead Lake" (no date but 1929), both RG1, PO, S15, B30, F "Biological Station Reports—1908, 1914, 1928–1929."

19. Robert T. Young, "The Life of Flathead Lake, Montana," in *Ecological Monographs* 5 (April 1935), 61–163, MJE, S5, B31, F35.

20. Morton J. Elrod to Robert H. Hill, State Game Warden, 10 May 1930, enclosing an accounting summary, a plan for the Missoula River issue, and a proposal to study lakes in Glacier National Park, MJE, S3, B6, F10. See also Morton J. Elrod to Robert H. Hill, 25 July 1930; and Robert H. Hill to Morton J. Elrod, 29 July 1930, both MJE, S2, B3, F2.

21. Elrod, "Recommendations" (no date but 1930), MJE, S5, B32, F10.

22. Elrod, "Vertebrates of Flathead Lake," 1929–30, RG1, PO, S15, B30, F "Biological Station Reports—1908, 1914, 1928–1929." On the salmon, see "Sockeye salmon," at http://en.wikipedia.org/wiki/Sockeye_salmon#Kunimasu.

23. Bonnie Ellis, "Long-Term Effects of Nonnative Species Introduction to Flathead Lake, Montana," *Montana Professor* 22, no. 2 (Fall 2012), 8–10; and "How Non-Native Shrimp Transformed the Ecosystem at Montana's Flathead Lake," at http://newwest.net/topic/article/how_non_native_shrimp_ transformed _the_ecosystem_at_montanas_flathead_lake/C41/L41/.

24. Chancellor Melvin A. Brannon to Morton J. Elrod, 8 May 1930, MJE, S3, B6, F10.

25. Floyd L. Smith, Editor, *Montana Wildlife*, to Morton J. Elrod, 24 November 1930, MJE, S3, B8, F1. See "Arctic Grayling—Thymallus arcticus," at http://fieldguide.mt.gov/detail_AFCHA07010.aspx. Morton J. Elrod to Byron DeForest, 3 March 1931, RG1, PO, S15, B30, F "Biological Station, 1930–1944."

26. "Whitefish Put in Skidoo Bay," *Montana Standard* (22 March 1931), MJE, S6, B34, F19.

27. Morton J. Elrod to Billy, 28 February 1929, MJE, S4, B14, F14.

28. Robert H. Hill to William G. Ferguson, 6 March 1929, MJE, S4, B14, F14.

29. W. J. Thompson to Morton J. Elrod, 30 March 1931, MJE, S4, B14, F16.

30. A. S. Hazard to Morton J. Elrod, 1 July 1932, MJE, S4, B14, F17.

31. Campus Development Committee, Minutes, 8 April 1929, RG1, PO, S6, B42, F "Campus Development, 1922–44."

32. "Honor Dr. Elrod," *Montana Wildlife* (May 1929), 13, MJE, S6, B34, F18.

33. C. H. Clapp to Chancellor M. A. Brannon, 2 February 1933; and Morton J. Elrod to President Clapp, 2 February 1933, both MJE, S3, B5, F; related letters, RG1, PO, S15, B30, F "Bio Station, 1930–1944"; and Beard, "Morton J. Elrod," Section 4.

34. Morton J. Elrod, "Biological Station," 14 September 1931, MJE, S5, B31, F1.

35. See RG1,PO, S15, B30, F "Biological Station, 1930–1944" and F "Biological Station, 1945–1955." Also Beard, "Morton J. Elrod," Section 4; and Giltner, "Montana State University," 15 February 1939, 4, RG1, PO, S15, B26, F "ARCHIVES: Informational."

36. Beard, "Morton J. Elrod," Section 4; President G. Finlay Simmons to Ray N. Shannon, Chairman, TRS, 8 August 1940, RG30, HR, F "Elrod, M. J."; and President James A. McCain to Gilbert J. Heyfron, 31 March 1950, RG1, PO, S15, B35, F "Law, 1935–1955."

37. Morton J. Elrod to Mrs. Helen Burn, 7 January 1933, MJE, S2, B2, F5.

38. Elrod, "The American University" (1916–20), MJE, S3, B4, F9.

39. Herbert L. Eastlick to Morton J. Elrod, 16 May 1933, MJE, S2, B2, F5.

40. J. Habeck, message of March 2013, in Jim Habeck, "Outline of Mary Elrod Ferguson's Timeline," draft, "Notes and Commentary on Mary Elrod Ferguson: Dr. James R. Habeck, January 2007," copy in author's possession; and Charles Adams to Morton J. Elrod, 12 September 1932, MJE, S2, B3, F5.

41. "Biographical Note," in "Guide to the Mary Elrod Ferguson Papers, 1898–1975," at http://nwda-db.wsulibs.wsu/findaid/ark:/80444/xv70455; and "Staff Recommendation," RG30, HR, F "Ferguson, Mary E. (Mrs.)."

42. Harriet Rankin Sedman, Dean of Women, State University, to Dean Sarah M. Sturtevant, Teachers College, 16 October 1933, MJE, S2, B3, F6; "Staff Recommendation," RG30, HR, F "Ferguson, Mary E. (Mrs.)"; and Morton J. Elrod to Mary, 29 April 1934, MJE, S2, B3, F7.

43. Morton J. Elrod to Mary, 30 May and 2 June 1934; and Harriet Rankin Sedman, Dean of Women, to President G. H. Vande Bogart (no date, but attached to Elrod's letter of 30 May 1934), all MJE, S2, B3, F8; and C. H. Clapp, President, to President Homer L. Schantz, 6 June 1934, RG30, HR, F "Ferguson, Mary E. (Mrs.)."

44. C. H. Clapp, President, "To the Members of the Faculty," 31 May 1933, with "Curricular Revision" attached, RG1, PO, S15, B35, F "Humanities, Division of."

45. H. H. Swain to R. L. Collins, Oregon, 25 July 1934, RG1, PO, S5, B47, F "Salaries, 1918–1944."

46. Morton J. Elrod, "Committee on Budget and Policy" Report (no date but Spring 1933); "Campus Development Committee" Report, 1933; "Committee on Museum" Report, 1933; "Committee on Service" Report, 1933; and "Committees of the Faculty," 1933–34, all MJE, S5, B23, F3. See also Clapp, "Narrative," ch. 7; and Merriam, *History,* ch. 6–7.

47. Clapp, "Narrative," ch. 7, 30–31; Elrod, "Committee on Budget and Policy" Report, 1933, MJE, S5, B23, F3; and President Charles H. Clapp to President S. E. Davis, 24 March 1933, RG1, PO, S5, B47, F "Salaries, 1918–1944."

48. Curriculum Committee Minutes, 9 December 1930, RG1, PO, S7, B14, F "Minutes, Curriculum Committee, 1928–1954." See also Merriam, *History,* 60–82.

49. Clapp's comments in Curriculum Committee Minutes, 9 December 1930, RG1, PO, S7, B14, F "Minutes, Curriculum Committee, 1928–1954."

50. Merriam, *History,* 71–72.

51. Elrod, "Campus Development Committee" Report, 1933, MJE, S5, B23, F3; Morton J. Elrod to President C. H. Clapp, 12 April 1933, MJE, S5, B28, F15; and T. G. Swearingen, "Campus Projects for WPA Improvements" (no date but penciled 6/15/38), RG1, PO, S6, B42, F "Campus Development, 1922–44."

52. *Philip R. Barber* vs. *State Board of Education et al.,* 9 July 1932, RG1, PO, S18, B21, F "Student Union"; Clapp, President's Report, 1932–33, 1933–34, UniPub, S2, B1929–1934, and B1934–1935, UniPub, S2, B1934–1938; and Merriam, *History,* ch. 6–7.

53. G. A. Jordan to Morton J. Elrod, 17 February 1933, MJE, S5, B24, F10.

54. The following analysis, except as noted, draws on Morton J. Elrod, "Aber Day Speech" (undated but 1932), MJE, S5, B26, F11; "Charter Day," draft (no date, but apparently a revision of the1916 talk written in the 1930s but not delivered); and "Charter Day Address," 1916, both MJE, S5, B26, F13; Elrod, "Charter Day—Information, For Members of the Faculty and Students," 18 February 1921; "To the Members of the Faculty," 15 February 1927; and "Proceedings of the Charter Day Committee . . . ," 17 February 1928, all RG1, PO, S4, B167, F "Charter Day—Informational." See also A. L. Stone, "An Address Delivered at the University of Montana Charter Day, 1914, RG1, PO, S19, B167, F "Charter Day Addresses 1912, 1914, 1918, 1919, 1920, 1921," on the first Aber Day in 1899.

55. Elrod, "Charter Day Address" (no date but revision of early 1930s), handwritten, urged a return to practices of many years ago, MJE, S5, B26, F13.

56. Elrod, "Many Changes," *Missoulian* (29 January 1922), 1, 5, MJE, S3, B7, F14.

57. On Aber, Morton J. Elrod, "Eulogy for Dr. Aber," 1919, MJE, S5, B26, F10.

58. On Ryman, M. J. Elrod, "Memorial" for J. H. T. Ryman, February 1927, who selected the original "trees and shrubbery" and served as a "member of the Campus Development Committee . . . at the time of his death," RG1, PO, S7, B29, F "RYMAN, J. H. T."; Whitlock to Speer, 29 December 1927, Ryman's bequest, RG1, PO, S5, B67, F "Ryman Scholarship."

59. For a similar perspective, see Lewis, *Excellence Without a Soul,* passim; and Menand, *Marketplace of Ideas,* passim.

60. Elrod to Emma, 29 July 1928, MJE, S2, B3, F1.

61. Charles C. Adams to Elrods, 13 June 1934, MJE, S2, B3, F8; also Helen Mims to Elrods, 12 February 1935, "glad I was to hear . . . that you, dear Dr. Elrod, were able to walk and speak," an erroneous impression, MJE, S2, B3, F9.

62. Mary Elrod Ferguson, typescript summary of Elrod's life, MJE, SI, B1, F1.

63. Ferguson to Thornton, 9 March 1954, MJE, S1 B1, F5.

64. See photograph of Morton J. and Emma Elrod in Glacier National Park in 1936 or 1937, he with a cane, markedly enfeebled, and she with a wrapped ankle, 87 herein, or in Digital Collections, image identifier umt015286, Toole Archives.

65. With no evidence that Elrod participated, see Morton J. Elrod, "Montana Teachers' Pension Law," *Inter-Mountain Educator* 10, no. 7 (March 1915), 24–25, for the 1915 "teachers' pension" of $50 a month for life to any teacher who paid $1 per month for at least 30 years of employment prior to retirement, also 10, no. 9 (May 1915) and others; and Merrill G. Burlingame and K. Ross Toole, *A History of Montana* (New York: Lewis Historical Publishing Co., Inc., 1957), vol. 2, 363–64, Montana's first Old Age Pension Law of 1923. In the early 1930s, Emma's brother died and included her in his will but the Depression devastated the estate.

66. "Staff Recommendation"; H. H. Swain to F. C. Scheuch, 12 July 1935; and F. C. Scheuch to H. H. Swain, 13 July 1935, all RG30, HR, F "Ferguson, Mary E. (Mrs.)."

67. Vice President R. H. Jesse, "Confidential Memorandum for the President concerning the museum personnel, referring to the letter of Dr. Phillips dated 3/1/52, received 4/25/52," 3 June 1952, RG1, PO, S15, B36, F "Museum: Northwest Historical Collection"; and Habeck, "Notes and Commentary," copy in author's possession.

68. For $400 annually, J. P. Rowe, Chairman, "The Budget and Policy Committee," 9 February 1935; and J. P. Rowe, Chairman, "Budget and Policy Committee," 13–14 March 1935, listing for 1935–36, both RG1, PO, S4, B42, F "Budget and Policy to 1936"; and for the denial, see H. H. Swain, Executive Secretary to the Board of Education, to Professor J. P. Rowe, 22 April 1935, RG1, PO, S4, B42, F "Budget and Policy Committee, 1936–41."

69. Ferguson, typescript summary of Elrod's life, MJE, SI, B1, F1. Habeck, "Notes and Commentary," copy in author's possession; President, State College of Washington, to Morton J. Elrod, 21 October 1938, MJE, S2, B3, F9; on Emma's health, Elrod to Burn, 7 January 1933, MJE, S2, B2, F5; to Emma, 28 March 1922, MJE, S2, B2, F7; and The Wahlins to Morton J. Elrod, University of Arkansas, 17 January 1933, MJE, S2, B3, F6.

70. Commencement, 6 June 1938, MJE, S2, B3, F9; and *Montana State University News Bulletin* 1, no. 3 (May 1938), 1–4, MJE, S6, B34, F20.

71. On TRS, see various letters and memoranda, RG1, PO, S5, B6, F "Retirement . . . to 1939," and F "Teachers Retirement System, 1940–1953."

72. Simmons to Shannon, 8 August 1940; and Mary Elrod Ferguson to TRS Board, 1939 and 1940, RG30, HR, F "Elrod, M.J."

73. See Simmons to Shannon, 8 August 1940; for handwritten note on the Simmons letter, 16 September 1940, "Approved for Dr. Elrod. Pres Simmons notified Mrs. Ferguson," RG30, HR, F "Elrod, M.J."

74. R. W. Harper, Executive Secretary, TRS, to Morton J. Elrod, 1 September 1947, MJE, S2, B3, F10. On Shope, see "Irvin (Shorty) Shope," *Cowboy Artists of America,* at http://www.caamuseum.com/members/deceased.irvin _shorty_shope.html.

75. "Staff Recommendation"; G. F. Simmons to M. E. Ferguson, 21 April 1937; and TRS Form, "Statement of Teaching Experience," 30 October 1937 with handwritten note on the back, all RG30, HR, F "Ferguson, Mary E. (Mrs.)."

76. Jesse, "Confidential Memorandum, "3 June 1952, RG1, PO, S15, B36, F "Museum: Northwest Historical Collection"; and Habeck, "Notes and Commentary," copy in author's possession. See also G. Finlay Simmons, President, to Mrs. Mary Elrod Ferguson, Assistant Dean of Women, 2 November 1940; and Mary Elrod Ferguson, "Acting Dean of Women," to President George Finlay Simmons, November 1940, both RG30, HR, F "Ferguson, Mary E. (Mrs.)."

77. Merriam, *History,* ch. 7; and documents relating to Simmons's troubles, RG1, PO, S4, B167, F "State Board Hearing, 1940."

78. James A. McCain, President, "Memorandum of Conversation with Mrs. Mary Elrod Ferguson," 13 March 1946; also President J. A. McCain to Dean J. E. Miller, Dean James L. C. Ford, Dr. Paul C. Phillips, and Mrs. Mary Elrod Ferguson, 9 October 1946; Paul C. Phillips to President J. A. McCain, 7 October 1946; and Paul C. Phillips to Vice President R. H. Jesse, 16 March 1953, all RG30, HR, F "Ferguson, Mary E. (Mrs.)"; and Habeck, "Notes and Commentary," copy in author's possession. See also Paul C. Phillips to Dean J. E. Miller, 5 September 1946, RG1, PO, S15, B36, F "Museum: Northwest Historical Collection."

79. Jesse, "Confidential Statement," 3 June 1952, RG1, PO, S15, B36, F "Museum: Northwest Historical Collection."

80. Carl McFarland to Mrs. Mary Elrod Ferguson, 21 March 1957; M. E. Ferguson to Carl McFarland, President, 25 May 1954, 7 April 1954, and 1 March 1957; A. S. Merrill, Vice President, to Mrs. Mary Elrod Ferguson, 11 April 1956; Carl McFarland, President, "Memorandum to Mrs. Ferguson," 13 March 1954 and 22 June 1954; and Paul C. Phillips to Vice President R. H. Jesse, 16 March 1953, with a note attached from Lucille Armsby, the president's secretary, all RG30, HR, F "Ferguson, Mary E. (Mrs.)."

81. Habeck, "Notes and Commentary," copy in author's possession. In a private conversation in 2013, Director Jack Stanford reported Elrod's cameras and a few other items at the Biological Station.

82. Campus Development Committee Minutes, 25 April 1935, RG1, PO, S6, B42, F "Campus Development, 1922–44."

83. Campus Development Committee Minutes, 20 April 1939 and 8 October 1940; and T. G. Swearingen to President Simmons, 5 May 1939, both RG1, PO, S6, B42, F "Campus Development, 1922–44." Also Dean C. E. Mollett to President George Finlay Simmons, 28 April 1937; and T. G. Swearingen, Maintenance Engineer, to President George Finlay Simmons, 10 February 1938, both RG1, PO, S15, B37, F "School of Pharmacy, 1932–1942"; and "Audubon Group Will Honor Dr. Morton J. Elrod," *Missoulian* (14 April 1942), MJE, S6, B34, F23.

84. Emmert to Elrod, 15 April 1952, and Mary E. Ferguson to J. W. Emmert, 22 May 1952, both MJE, S2, B3, F11.

85. "Kalispell School Is Dedicated For MSU's Dr. Morton J. Elrod," *Missoulian* (29 March 1952), MJE, S6, B34, F21; and Habeck, "Notes and Commentary," copy in author's possession.

86. "Three Buildings to Be Renamed," *Missoulian* (8 February 1956), MJE, S6, B34, F23.

87. Habeck, "Notes and Commentary," esp. "Interview #8," copy in author's possession.

88. See résumés in the Elrod papers, exemplified in MJE, S2, B3, F10.

89. Isaiah Bowman, Director, American Geographical Society, to Morton J. Elrod, 1916, handwritten "Told him I could not do it," MJE, S2, B2, F2.

90. Victor E. Shelford, Ecological Society of America, to Morton J. Elrod, 7 July 1917, handwritten "ans. July 20," MJE, S2, B2, F3.

91. R. Rathun, Assistant Secretary, Smithsonian Institution, to M. J. Elrod, 31 August 1900; H. A. Pilsby, The Academy of Natural Sciences of Philadelphia, to Morton J. Elrod, 19 and 30 July 1900, all MJE, S2, B1, F9; and Morton J. Elrod to Malchoir (indecipherable), 14 March 1915, weather conditions in late November for the football game between Montana and Syracuse, MJE, S2, B1, F12.

92. Sydelle M. Haskell, A.N. Marquis Co., to Morton J. Elrod, 21 February 1949, MJE, S2, B3, F10.

Postscript

1. Morton J. Elrod to Missoula City Council, 25 March 1915, MJE, S2, B1, F12.

2. Clapp, "Narrative," ch. 5, 17–18.

3. Elrod, "Many Changes," *Missoulian* (29 January 1922), 1, 5, MJE, S3, B7, F14; and Merriam, *History*, 182.

Selected Bibliography

Primary Sources

Archives and Special Collections. Digital Collections, K. Ross Toole Archives. Maureen and Mike Mansfield Library. University of Montana.

Clyde Augustus Duniway Papers, Relating to the University of Montana, 1908–1912. Microfilm (not including Reel 1). K. Ross Toole Archives. Maureen and Mike Mansfield Library. University of Montana.

Digitized Newspapers. Montana Historical Society. Helena, Montana.

George Bird Grinnell Papers. Microfilm. Maureen and Mike Mansfield Library. University of Montana.

Human Resource Records. Record Group no. 30. K. Ross Toole Archives. Maureen and Mike Mansfield Library. University of Montana.

Inter-Mountain Educator. K. Ross Toole Archives. Maureen and Mike Mansfield Library. University of Montana.

Joseph M. Dixon Papers. Microfilm. K. Ross Toole Archives. Maureen and Mike Mansfield Library. University of Montana.

Missoulian. Microfilm. Maureen and Mike Mansfield Library. University of Montana.

Morton J. Elrod Papers, no. 486. K. Ross Toole Archives. Maureen and Mike Mansfield Library. University of Montana.

Record Group no. 1. The Office of the President. K. Ross Toole Archives. Maureen and Mike Mansfield Library. University of Montana.

"Smoke Fumes" Scrap Book. MS 753. Scrapbook 4. K. Ross Toole Archives. Maureen and Mike Mansfield Library. University of Montana.

State Board of Education. Minutes. Maureen and Mike Mansfield Library. University of Montana.

University Publications. No. 2. President's Reports, bound. K. Ross Toole Archives. Maureen and Mike Mansfield Library. University of Montana.

Weekly Missoulian. Microfilm. Maureen and Mike Mansfield Library. University of Montana.

Secondary Sources

Alt, David D. *Glacial Lake Missoula: And Its Humongous Floods*. Missoula, MT: Mountain Press Publishing Co., 2001.

American Bison Society. *Annual Report, 1905–1907*. http://archive.org/stream /annualreportofambs00amer#page/n29/mode/2up.

———. *Second Annual Report*. 1909. http://www.forgottenbooks.com/readbook /Annual_Report_of_the_American_Bison_Society_1908_1000510071 #119.

Ashby, Christopher S. "Blackfeet Agreement of 1895 and Glacier National Park: A case history." 1985. *Theses, Dissertations, Professional Papers*, 1684. http://scholarworks.umt.edu/cgi/viewcontent.cgi?article=2703& context=etd.

Athearn, Robert G. "The Tin Can Tourists' West." In *Montana and the West: Essays in Honor of K. Ross Toole*, edited by Rex C. Myers and Harry W. Fritz. Boulder, CO: Pruett Publishing Company, 1984. 105–21.

"Audubon Group Will Honor Dr. Morton J. Elrod." *Missoulian*. (14 April 1942).

Bloom, Allan. *The Closing of the American Mind*. New York: Simon and Schuster, 1967.

Bottomly-O'Looney, Jennifer, and Kirby Lambert. *Montana's Charlie Russell Art in the Collection of the Montana Historical Society*. Helena, MT: Montana Historical Society, 2014.

Bosworth, Brendon. "How Non-Native Shrimp Transformed the Ecosystem at Montana's Flathead Lake." *New West*. 21 January 2011. http://newwest .net/topic/article/how_non_native_shrimp_transformed_the_eco system_at_montanas_flathead_lake/C41/L41/.

Buchholz, C. W. *Man in Glacier*. http://www.nps.gov/history/history/online _books/glac2/.

Burlingame, Merrill G., and K. Ross Toole. *A History of Montana*. New York: Lewis Historical Publishing, 1957. 2 vols.

Carr, Edward Hallett. *What is History?* New York: Vantage Books, 1961.

Chomsky, Noam. *New Horizons in the Study of Language and Mind*. Cambridge: Cambridge University Press, 2000.

Clapp, Mary Brennan. "Narrative of Montana State University: 1893–1935." 1958. Typescript copy in author's possession, another in K. Ross Toole Archives and Special Collections, Maureen and Mike Mansfield Library, University of Montana.

Court of Appeals of the United States. "Bliss v. Washoe Copper Co., et al." In *Federal Reporter* 186. St. Paul: West Publishing, 1911. 789–828. Available at https://books.google.com/books?id=HyQ4AAAAIAAJ&lpg= PA801&dq=Bliss%20v.Washoe%20Copper%20Smelter&pg=PA801 #v=onepage&q=Bliss%20v.Washoe%20Copper%20Smelter&f=false.

Deibler, Frederick S. "Committee on Academic Freedom: Statement on the Case of Professor Louis Levine of the University of Montana." AAUP.

Bulletin of the American Association of University Professors. Reports of Actions on Carnegie Foundation Proposals; University of Montana; Bethany College; List of Committees. Boston: AAUP, March 1919. 13–25.

Deibler, F. S. "Report on the University of Montana." *Bulletin of the American Association of University Professors* 10, no. 3 (March 1924): 154–62.

Dennison, George M. "Higher Education in Montana, 1950–1993." *Montana: The Magazine of Western History* 44, no. 2 (Spring 1994): 65–72.

Desrochers, Donna M., and Rita Kirshstein. "Labor Intensive or Labor Expensive? Changing Staffing and Compensation Patterns in High Education." Delta Cost Project, American Institutes for Research, February 2014.

"Did Not Consider Alkali. . . ." *The Butte Inter Mountain* 4 (April 1906): 1, 5.

Diettert, Gerald A. *Grinnell's Glacier: George Bird Grinnell and Glacier National Park.* Missoula: Mountain Press, 1992.

Dimsdale, Thomas J. *The Vigilantes of Montana or Popular Justice in the Rocky Mountains.* 3rd ed. Helena, MT: State Publishing, 1915.

"Dr. Morton J. Elrod is Re-Elected." *Missoulian.* 15 September 1908.

Ellis, Bonnie. "Long-Term Effects of Nonnative Species Introduction to Flathead Lake, Montana." *Montana Professor* 22, no. 2 (Fall 2012): 8–10.

Elrod, Emma H. "Flathead Indian Reservation." In *The Woman's Souvenir of Missoula Montana,* edited by Elizabeth L. Mills. Missoula: Ladies of the Christian Church, 1910. 28.

Elrod, Morton. "An Electric Storm in Montana." *Youth's Companion* (2 November 1911) and *Kaimin* (1911).

"Elrod, Morton J., 1863–1953." Prototype Research History Tool. Institute for Advanced Technology in the Humanities. http://socialarchive.iath .virginia.edu/ark:/99166/w60g3nb4.

Elrod, Morton J. "A Camp at Mt. Lolo." *The Outdoor World.* 1899. 65–68.

———. "Among the Kootenais." *Rocky Mountain Magazine* 3, nos. 3–4 (November–December 1901): 171–85.

———. "Among the Rockies." *The Wesleyan Argus* 1, no. 2 (28 September 1894): 5–7; no. 5 (12 November 1894): 5–8; no. 10 (28 January 1895): 5–9; no. 16 (28 April 1895): 6–9.

———. "Ascent of Mount Lolo." *Illinois Wesleyan Magazine.* 1893: 322–28.

———. "Beauties of the Mission Range," *Rocky Mountain Magazine* 2, no. 2 (April 1901): 623–31.

———. *Biological Reconnaissance in the Vicinity of Flathead Lake.* Reprint Edition, Cornell University Library, Digital Collections, OH 105.M9E 48, no date. Originally *Bulletin University of Montana* Biological Series 3, no. 10 (1902).

———. "Blackfoot Ice Mass, Glacier Park, Receding." *Missoulian.* 21 March 1926.

———. *Butterflies of Montana: With Keys for Determination of Species.* Missoula: University of Montana, 1906.

———. *The College, Past and Present.* Bloomington. IL: University Press, 1899.

———. "Conquering Glacier Park Peaks." *Missoulian.* 3 February 1929.

———. "Consolidation of the State Institutions." *Inter-Mountain Educator* 9, no. 9 (May 1914): 27–30.

———. "Consolidation of the State Institutions." *Inter-Mountain Educator* 9, no. 10 (June 1914): 18–21, 24–25.

———. "Consolidation of State Institutions." *Inter-Mountain Educator* 10, no. 1 (September 1914): 25–26.

———. "Consolidation of the University, the Agricultural College and the School of Mines." *Inter-Mountain Educator* 10, no. 2 (October 1914): 7–20, 25–26.

———. "Consolidation and Women's Suffrage." *Inter-Mountain Educator* 10, no. 4 (November 1914): 24.

———. "Dr. Louis Levine Reinstated." *Inter-Mountain Educator* 14, no. 8 (April 1919): 41–42.

———. "Educational Building in Glacier National Park Would Give Visitors Better Understanding of Wonderland, Says Elrod." *Great Falls Tribune.* 2 September 1923.

———. "Elrod Describes Grinnell Glacier." *Missoulian.* 13 May 1923.

———. *Elrod's Guide and Book of Information of Glacier National Park.* 2nd ed., revised and enlarged. Missoula: Morton J. Elrod, 1930.

———. "Evolution and Religion Do Not Conflict, Say Teachers." *Missoulian.* 31 May 1925.

———. "The Flathead Buffalo Range: A Report to the American Bison Society . . . ." *Annual Report, 1905–1907.* American Bison Society, 1908. 15–49.

———. "Following Old Trails." *Inter-Mountain Educator* 9, no. 6 (February 1914): 25–26.

———. "Garden Wall in Glacier National Park Presents a Scene of Great Beauty Unexcelled." *Missoulian.* 25 June 1911.

———. "Glacier National Park." *Encyclopedia Americana* (1919); also *The Missoula Sentinel.* 26 April 1919.

———. "Guides of Glacier Park Are Real Men, Have Arduous Tasks." *Missoulian.* 24 January 1926.

———. "Iceberg Lake." *Mountaineer.* n.d., but 1910–1915. 43–51.

———. "Inter-Mountain Educator." *Montana Education* 1, no. 3 (November 1924): 15–16.

———. "Many Changes in the City Are Told by Elrod." *Missoulian.* 29 January 1922.

———. "Minimum Essentials for High School Zoology." *Inter-Mountain Educator* 16, no. 7 (March 1921): 303–306.

———. "M. J. Elrod Returns From Glacier Park." *Missoulian.* 5 September 1910.

———. "The Montana Bison Range." *Journal of Mammalogy* 7, no. 1 (February 1926): 45–48.

——. "Montana Professor Suspended." *Inter-Mountain Educator* 14, no. 6 (February 1919): 24–26.

——. "Montana's New National Park Is Grand, A Priceless Pleasure Ground For All." *Missoulian.* 24 July 1910.

——. "Montana State Board of Education Acts on Duplication of Courses." *Inter-Mountain Educator* 9, no. 1 (September 1913): 26–27.

——. "Montana Teachers' Pension Law." *Inter-Mountain Educator* 10, no. 7 (March 1915): 24–25.

——. "Montana University and Agricultural College Presidencies are Vacant." *Inter-Mountain Educator* 10, no. 10 (June 1915): 5.

——. "Mountain Climbing in Glacier." *Missoulian.* 27 January 1929.

——. "Parasitic Copepoda Found on Fins and Gills of Bull Trout." *Proceedings of the United States Museum* 47. 1915.

——. "Passing of the Pablo Buffalo Herd." *Shields' Magazine* 12, no. 2 (February 1911): 35–41.

——. "Play Fair, Boys." *Inter-Mountain Educator* 17, no. 5 (January 1922): 216.

——. "Professor Fisher." *Inter-Mountain Educator* 17, no. 2 (October 1921): 73–74.

——. "Proposed Glacier National Park." With a line drawn through "Proposed." No date but written before 11 May 1910: MJE, SIV, B18, F15.

——. *Some Lakes of Glacier Park.* Washington, DC: Department of the Interior, 1912.

——. *Vacation in Montana.* Bloomington, IL: University Press, 1899.

——. *Views of the Mission Mountains . . . Flathead Lake and Valley, Montana.* Missoula: Morton J. Elrod, 1908.

——. "Where the Red Man Wrote His Story on the Rocks and Left the Record of Great Salish Triumphs." *Missoulian.* 1 March 1908.

Fish and Game Commission. *Montana Wildlife* 2, no. 1 (June 1929): 3–15.

Gale, Gil. "Fire Ferocity: Wildfires of the Past." *Montana Naturalist* (Winter 2014–2015): 9.

Garvin, Ellen. "And in the Beginning. . . ." Senior Practice Laboratory. School of Journalism, the State University of the University of Montana. Missoula: 18 May 1925. Copy in author's possession.

Greenleaf, Walter J. "The Land-Grant College as It Functions Today." U.S. Office of Education. *School and Society* 41 (29 June 1935): 855–59.

Grinnell, George Bird. "Crown of the Continent." *Century Magazine* 62, no. 76 (September 1901): 660–72.

Gutfeld, Arnon. *Montana's Agony: Years of War and Hysteria, 1917–1921.* Gainesville: University Presses of Florida, 1979.

——. "The Levine Affair: A Case Study in Academic Freedom." *Pacific Historical Review* 39, no. 1 (February 1970): 19–37.

Guthrie, C. W. *The Great Northern Railway and Glacier National Park.* Helena, MT: Far West, 2004.

Habeck, Jim. "Can Botanists Be Bought?" *Kelseya* (Summer 2006): http://www
.mtnativeplants.org/filelib/75.pdf.

Hamilton, Neil W. *Academic Ethics: Problems and Materials on Professional
Conduct and Shared Governance.* Washington, DC: American Council
on Education and Praeger Publishers, 2002.

Hanson, James E. *Democracy's College in the Centennial State: A History of
Colorado State University.* Fort Collins: Colorado State University,
1977.

Harper, Andrew C. "Conceiving Nature: The Creation of Montana's Glacier
National Park," *Montana: The Magazine of Western History* 60, no. 2
(Summer 2010): 3–24, 91–94.

Haynes, Robert F. "About Dr. Morton Elrod, Park's First Naturalist." *Hungry
Horse News* 14 (December 1956).

"Honor Dr. Elrod." *Montana Wildlife.* May 1929. 13.

Howard, Joseph Kinsey. *Montana: High, Wide, and Handsome.* Reprint ed. Lincoln and London: University of Nebraska Press, 1983.

Hurst, James Willard. *Law and the Conditions of Freedom in the Nineteenth-
Century United States.* Madison: University of Wisconsin Press, 1956.

Isaacson, Walter. *Einstein: His Life and Universe.* New York: Simon and Schuster, 2007.

"Kalispell School Is Dedicated For MSU's Dr. Morton J. Elrod." *Missoulian.* 29
March 1952.

Karlin, Jules A. "Conflict and Crisis in University Politics." *Montana: The Magazine of Western History* 36, no. 2 (Summer 1985): 48–61.

Karlin, Jules Alexander. *Joseph M. Dixon of Montana.* Missoula: University of
Montana Press, 1974. 2 vols.

Katznelson, Ira. *Fear Itself: The New Deal and the Origins of Our Time.* New
York: Liveright Publishing Company, 2013.

"The Keeney Case: Big Business, Higher Education, and Organized Labor,
Report of an Investigation." Chicago: American Federation of Teachers,
1939. http://www.worldcat.org/title/keeney-case-big-business-higher
-education-and-organized-labor-report-of-an-investigation/oclc/134
91059.

Kerstetter, Tulli. "J. W. Blankinship." http://www.montana.edu/mlavin/herb
/jwb.htm.

Knoll, Robert E. *Prairie University: A History of the University of Nebraska.* Lincoln: University of Nebraska Press, 1995.

Kofoid, Charles A. "Academic Freedom and Academic Tenure: Report of the
Committee of Inquiry Concerning Charges of Violation of Academic
Freedom, Involving the Dismissal of the President and Three Mem-
bers of the Faculty at The University of Montana." *Bulletin of the
American Association of University Professors* 3, no. 5, part 2 (May
1917): 3–52.

Leonard, Thomas C. *Illiberal Reformers: Race, Eugenics, and American Economics in the Progressive Era.* Princeton, NJ: Princeton University Press, 2016.

Levine, Louis. *The Taxation of Mines in Montana.* Reprint ed. Boulder: University of Colorado BiblioLife, n.d. Originally published in 1919.

Lewis, Harry R. *Excellence Without a Soul: How a Great University Forgot Education.* New York: Public Affairs, 2006.

MacMillan, Donald. "A History of the Struggle to Abate Air Pollution From Copper Smelters of the Far West, 1885–1933." PhD diss., University of Montana, 1973.

———. *Smoke Wars.* Helena: Montana Historical Society Press, 2000.

Malone, Michael P. *Battle for Butte: Mining and Politics on the Northern Frontier, 1864–1906.* Seattle: University of Washington Press, 1981.

———. "The Montana University System: The First Half Century." *Montana: The Magazine of Western History* 44, no. 2 (Spring 1994): 60–64.

Malone, Michael P., and Richard B. Roeder. *Montana: A History of Two Centuries.* Seattle and London: University of Washington Press, 1988.

Marx, Leo. *The Machine in the Garden: Technology and the Pastoral Ideal in America.* London and New York: Oxford University Press, 1964.

McGrath, Earl J. *The Graduate School and the Decline of Liberal Education.* New York: Institute of Higher Education, 1959.

McGrath, Earl J., and Charles H. Russell. *Are Liberal Arts Colleges Becoming Professional Schools.* New York: Institute of Higher Education, 1958.

Menand, Louis. *The Marketplace of Ideas.* New York and London: W. W. Norton, 2010.

Merriam, H. G. *The University of Montana: A History.* Missoula: University of Montana Press, 1970.

Miller, Dan, and Stan Cohen. *The Big Burn: The Northwest's Great Forest Fire of 1910.* Missoula: Pictorial Histories Publishing, 1978.

Montana State University News Bulletin 1, no. 3 (May 1938): 1–4.

Morand, Ann. *Your Friend, C. M. Russell: The C. M. Russell Museum Collection of Illustrated Letters.* Great Falls, MT: C. M. Russell Museum, 2008.

National Park Service. *Glacier: Through the Years in Glacier National Park.* http://www.nps.gov/history/history/online_books/glac/chap3.htm.

O'Donnell, Richard F. "Higher Education's Faculty Productivity Gap: The Cost to Students, Parents, and Taxpayers." 2011. http://gaia.pge.utexas.edu/papers/Higher%20Eds%20Faculty%20Productivity%20Gap.pdf.

Peterson, Larry Len. *The Call of the Mountains: The Artists of Glacier Park.* Tucson, AZ: Settlers West Galleries, 2002.

Piketty, Thomas. *Capital in the Twenty-First Century.* Cambridge, MA, and London: Belknap Press of Harvard University Press, 2014.

Rice, Richard E., and George B. Kauffman. "William Draper Harkins: An Early Environmental Chemist in Montana, 1900–1912." *Bulletin for the History of Chemistry* 20 (1997): 60–67.

Rile, Judith A. "The Changing Role of the President in Higher Education." 2001. http://www.newfoundations.com/OrgTheory/Rile721.html.

Rydell, Robert E., Jeffrey Safford, and Pierce Mullen. *In the People's Interest: A Centennial History of Montana State University.* Bozeman: Montana State University, 1993.

Segerstralle, Ullica. *Nature's Oracle: The Life and Work of W. D. Hamilton.* Oxford, UK: Oxford University Press, 2013.

Smith, Burton. "The Politics of Allotment: The Flathead Reservation as a Test Case." *Pacific Northwest Quarterly* 70, no. 3 (July 1979): 130–40.

"Smoke Not Wholly to Blame. . . ." *The Butte Inter Mountain* 3 (April 1906): 1, 8.

Speer, J. B. "A Bird's-Eye View of the Organization of One University." *Journal of Higher Education* 2, no. 9 (December 1933): 461–67.

"State Board Reaffirms Previous Action as to President Duniway." *Missoulian.* 2 April 1912.

Stearns, Sheila MacDonald. "The Arthur Fisher Case." Master's thesis, the University of Montana, Missoula, 1969.

Stone, Arthur L. *Following Old Trails.* Missoula: Morton John Elrod, 1913.

Swain, R. E., and W. D. Harkins. "Papers on smelter smoke." *Journal of American Chemical Society* 29, no. 4 (April 1907) and 30, no. 6 (June 1908). https://archive.org/stream/papersonsmelters00harkrich/paperson smelters00harkrich_djvu.txt.

"Three Buildings to Be Renamed." *Missoulian.* 8 February 1956.

Vaughn, Richard. "To the Ice: George Bird Grinnell's 1887 Ascent of Grinnell Glacier." 2010. http://www.repository.law.indiana.edu/facpub/747/.

Wiebe, Robert H. *The Search for Order: 1977–1920.* New York: Hill and Wang, 1967.

Wootton, David. *The Invention of Science: A New History of the Scientific Revolution.* New York: HarperCollins, 2015.

Work, Clement P. *Darkest Before Dawn: Sedition and Free Speech in the American West.* Albuquerque: University of New Mexico Press, 2005.

"Would not Venture An Opinion. . . ." *The Butte Inter Mountain* 5 (April 1906): 1, 10.

Young, Robert T. "The Life of Flathead Lake, Montana." *Ecological Monographs* 5 (April 1935): 61–163.

Zwick, Rebecca. "College Admission Testing." Arlington, VA: National Association for College Admission Counseling, 2007. www.nacacnet.org /media-center/standardizedtesting/documents/standardizedtesting whitepaper.pdf.

Index

Page numbers in *italics* refer to illustrations.